Annals of the ICRP

ICRP PUBLICATION 91

A Framework for Assessing the Impact of Ionising Radiation on Non-human Species

Editor
J. VALENTIN

PUBLISHED FOR

The International Commission on Radiological Protection

by

PERGAMON

CONTENTS

PERGAMON

ICRP Publication 91

Annals of the ICRP

Guest Editorial

THE COMMISSION'S POLICY ON THE ENVIRONMENT

The International Commission on Radiological Protection (ICRP)'s current position regarding protection of the environment is set out in its *Publication 60* (ICRP, 1991): '*The Commission believes that the standards of environmental control needed to protect man to the degree currently thought desirable will ensure that other species are not put at risk.*' Until now, the Commission has not published any recommendations as to how protection of the environment should be carried out. The Commission therefore set up a Task Group in the year 2000 to address this issue, and has recently adopted the present report from that Task Group. This report is directed to the Commission itself, and addresses the role that it could play in this important and developing area, building on the approach that has been developed for human protection and on the Commission's specific area of expertise, namely radiological protection.

At present, there are no internationally agreed criteria or policies that explicitly address protection of the environment from ionising radiation, although many international agreements and statutes call for protection against pollution generally, including radiation. The lack of both a policy and a technical basis for assessment, criteria, or standards that have been endorsed at an international level makes it difficult to determine or demonstrate whether or not the environment is adequately protected from potential impacts of radiation under different circumstances. The international development of an explicit assessment framework would support and provide transparency to the decision-making process.

The Commission's system for protection of human beings has indirectly provided a fairly good level of protection of the human habitat. The Commission's decision to develop a system to develop a framework for the assessment of radiation effects in non-human species has not been driven by any particular concern over environmental radiation hazards. It has rather been developed to fill a conceptual gap in radiological protection, and to clarify how the proposed framework can contribute to the attainment of society's goals of environmental protection by developing a protection policy based on scientific and ethical-philosophical principles.

The Commission's framework will be designed so that it is harmonised with its proposed approach for the protection of human beings. To achieve this, an agreed set of quantities and units, a set of reference dose models, reference dose-per-unit-intake data, and reference organisms will be developed. As a first step, a limited number of reference fauna and flora will be developed by the Commission, and others can then develop more area- and situation-specific approaches to assess and manage risks to non-human species. Two of the most important features of human radiological protection are the continuing development of the 'Reference Man' concept as the basis for estimating radiation effects, plus the new proposed concept of expressing different individual annual doses in terms of 'levels of concern', using natural background dose rates as a reference point.

The proposed system does not intend to set regulatory standards. The Commission rather recommends a framework that can be a practical tool to provide high-level advice and guidance, and help regulators and operators demonstrate compliance with existing legislation. In contrast to the unique position of the ICRP in relation to human radiological protection, from which it has played a major role in influencing legal frameworks and objectives at international and national levels, the subject of protection of other species is a more complex and multifaceted one, with many international and national environmental legislative frameworks and objectives already in place.

The Commission will therefore develop a small set of reference fauna and flora, plus their relevant databases—in a manner similar to that of Reference Man—to serve as a basis for the more fundamental understanding and interpretation of the relationships between exposure and dose and between dose and certain categories of effect for a few, clearly defined types of animals and plants. The magnitude of doses relating to these effects will be set out in a 'banded' fashion, such as the proposed derived consideration levels, in a manner similar to the levels of concern considered for human beings. Such a set of information could then serve as a basis from which national bodies could develop, as necessary, more applied and specific numerical approaches to the assessment and management of risks to non-human species as national needs and situations arise.

This concept of deriving such data sets for reference fauna and flora is therefore similar to that of the reference individual (Reference Man) used for human radiological protection, in that it is intended to act as a basis for many calculations and decisions. It is intended that each reference organism would serve as a *primary* point of reference for assessing risks to organisms with similar life cycles and exposure characteristics. More locally relevant information could be compiled for any other fauna and flora, but each such data set would then have to be shown to be related in some way to the reference organisms.

The Commission acknowledges the important work being done by other national and international bodies, and the need to work in consort with them to develop the field of environmental protection. A considerable challenge for ICRP will be that of integrating any approach to protection of the environment with that of the protection of human beings, bearing in mind that the latter is also the subject of a current, in-depth review.

It is necessary that a system for radiological protection of non-human organisms be harmonised with the principles for the radiological protection of humans. The objectives of a common approach to the radiological protection of non-human organisms might be to safeguard the environment by:

- preventing or reducing the frequency of effects likely to cause early mortality or reduced reproductive success in individual fauna and flora to a level where they would have a negligible impact on

- conservation of species, maintenance of biodiversity, or the health and status of natural habitats or communities.

ICRP can and is prepared to play the key role with respect to ionising radiation, both in advising on a common international approach, and in providing the basic interpretation of existing scientific information—and identifying where further research is necessary—in order for such a common approach to be delivered.

The Commission's system of protection has evolved over time as new evidence has become available and as our understanding of underlying mechanisms has increased. Consequently, the Commission's risk estimates have been revised regularly, and substantial revisions have been made at intervals of about 10–15 years. It is therefore likely that any system designed for the radiological protection of the environment would also take time to develop, and similarly be subject to revision as new information is obtained and experience gained in putting it into practice.

ROGER H CLARKE
LARS-ERIK HOLM

PREFACE

At its meeting in Vienna in May 2000, the Main Commission of the ICRP decided to set up a Task Group, reporting directly to the Commission, in order to produce a report on the protection of the environment. The Group was tasked with the aim of developing both a protection policy and suggesting a framework for environmental protection based on scientific and ethical-philosophical principles. The proposed concepts for such a protection framework contained in this document include provisions for protecting non-human species, rather than the entire biotic and abiotic components, and ecosystems of the environment. These concepts are expected to feed into the Main Commission's deliberations on how the ICRP should proceed concerning its policy on the protection of living organisms, and thereby also into the preparation of the ICRP's next recommendations.

The membership of the Task Group was as follows:

L.-E. Holm (Chairman), Sweden
R. Alexakhin, Russian Federation
R.J. Pentreath, UK
P. Strand, Norway
K. Shrader-Frechette, USA
P.-A. Thompson, Canada

The following persons served as corresponding members:

F. Brechignac, France
D. Cancio, Spain
S. Carroll, Greenpeace International
M.E. Clark, USA
S. Domotor, USA
F. Fry, UK
K. Fujimoto, Japan
N. Gentner, UNSCEAR
G. Hunter, European Commission
A. Janssens, European Commission
C.-M. Larsson, Sweden
I. Likhtarev, Ukraine
C. Mothersill, Ireland
C. Robinson, IAEA
S. Sadasivan, India
S. Saint-Pierre, France
R. Saxén, Finland
A. Shpyth, Canada
S. Sundell-Bergman, Sweden
D.S. Woodhead, UK
H. Yang, China
A. Zapantis, Australia

The Task Group would like to thank those organisations that made facilities available for Task Group meetings: the Swedish Radiation Protection Authority, the International Atomic Energy Agency, the Australian Office of the Supervising Scientist, and the Spanish Centre for Research on Energy, on the Environment and on Technology.

The report was adopted by the Main Commission at its meeting in Vienna, 24–25 January 2003.

PERGAMON

ICRP Publication 91

Annals of the ICRP

A framework for assessing the impact of ionising radiation on non-human species

ICRP Publication 91

Approved by the Commission in October 2002

Abstract-In its 1990 Recommendations, the ICRP indicated that it believed that the standards of environmental control needed to protect man to the degree currently thought desirable would ensure that other species are not put at risk. The ICRP considers that its system of radiological protection has provided a fairly good indirect protection of the human habitat. However, no internationally agreed criteria or policies explicitly address protection of the environment from ionising radiation, and it is difficult to determine or demonstrate whether or not the environment is adequately protected from potential impacts of radiation under different circumstances. The present report suggests a framework, based on scientific and ethical-philosophical principles, by which a policy for the protection of non-human species could be achieved. The primary purpose of developing such a framework is to fill a conceptual gap in radiological protection; it does not reflect any particular concern over environmental radiation hazards.

The proposed framework is designed to harmonise with the ICRP's approach to the protection of human beings, but does not intend to set regulatory standards. Instead, the proposed framework is intended to be a practical tool to provide high-level advice and guidance for regulators and operators. An agreed set of quantities and units, a set of reference dose models, reference dose-per-unit-intake (or unit exposure), and reference fauna and flora are required to serve as a basis for the more fundamental understanding and interpretation of the relationships between exposure and dose and between dose and certain categories of effect, for a few, clearly defined types of animals and plants. As a first step, a small set of reference fauna and flora with supporting databases will be developed by the ICRP. Others can then develop more area- and situation-specific approaches to assess and manage risks to non-human species.

Keywords: Environment; Radiation protection; Biocentric; Sustainable development; Reference model.

EXECUTIVE SUMMARY

In May 2000, the Main Commission of ICRP decided to set up a Task Group to advise on the development of a policy for the protection of the environment, and to suggest a framework—based on scientific and ethical-philosophical principles—by which it could be achieved. This was new ground for the Commission because it had previously considered exposures of other organisms to ionising radiation only in so far as they related to the protection of human beings. In contrast to the Commission's unique position in relation to human radiological protection, from which it has played a major role in influencing legal frameworks and objectives at international and national levels, the subject of environmental protection is a more complex and multifaceted one, with many international and national environmental legislative frameworks and objectives already in place.

The current and potential future role of the Commission with respect to protecting the environment, by way of an understanding of the effects of ionising radiation on animals and plants, has therefore been discussed against this background. It was concluded that the principal contribution that the Commission could make was that of providing broad policy and guidance—as it does for human radiological protection—by way of formulating recommendations and advice, supported by some key data sets and models. Indeed, it was considered essential that the Commission should develop a more comprehensive approach to the study of the effects on, and thus the protection of, all living matter with respect to ionising radiation, and that it should therefore develop its 'system of protection' to include both humans and other living things generally.

If such an approach were to be taken by ICRP, it is also clear that it would need to work in consort with other bodies. The relative roles of the United Nations Scientific Committee on the Effects of Atomic Radiations (UNSCEAR), the International Commission on Radiation Units and Measurements (ICRU), the International Atomic Energy Agency (IAEA), the Nuclear Energy Agency (NEA), the International Radiation Protection Association (IRPA), and the International Union of Radioecology (IUR) are therefore briefly discussed, as well as those of international bodies who have a need for, and would also play a role in, the practical achievement of environmental protection, including, again, the above organisations plus such bodies as The Oslo-Paris Convention for the Protection of the Marine Environment of the North East Atlantic (OSPAR), the European Union (EU) and so on. Such an approach by the ICRP would also have to be cast within the current ethical and social views of what constitutes environmental protection generally, and how such different views, and broadly agreed principles, help to define it.

The recent IAEA study (IAEA, 2002a) was considered to provide a sound basis upon which to proceed, by drawing together the current ethical views of relevance—anthropocentric, biocentric, and ecocentric—plus the 'principles' embodied in United Nations (UN) legislation of sustainable development, conservation of the natural world, and the need to maintain biological diversity. All of these concepts are also supported by the need to provide environmental justice, and to have respect for human dignity. These are all complex and inter-related issues, and they have been variously addressed at international level over the last three decades. Of particular importance has been the concept of sustainable development, including recognition of the need to protect all living resources. Such concepts have had a large impact globally since the 'Rio' Convention of 1992, and hence since the publication of the Commission's *Publication 60* in 1991. Similarly, approaches to the assessment and management of environmental risks are continually changing, and such changes will inevitably need to be reflected in the Commission's deliberations on its approach to the protection of non-human species.

If ICRP is to develop a more comprehensive approach to the protection of living matter, it also needs to reconsider existing databases and their interpretation. The majority of our information on the exposure to, and effects of, radiation has been derived in order to serve the needs of human radiological protection. Probably the most important first step to take is that of distinguishing between the manner in which radiation effects are expressed, at the level of the individual, in different types of animals and plants. For humans, the main concern has been that of safeguarding health by an understanding of the way in which effects can be characterised as being stochastic or deterministic. But there is insufficient knowledge to enable a similar approach to be considered for non-human species—with the possible exception of some mammals. It is therefore considered that a more useful approach would be to describe the effects of radiation on individuals in categories that would be relevant in an environmental context, such as causing early mortality (by any means), or some form of morbidity, or a decrease in reproductive success. The extent to which such effects might, in turn, have consequences for populations, communities, or even entire ecosystems would depend on a large number of factors, including not only the number of individuals variously exposed to radiation, but also many other factors unrelated to ionising radiation.

A considerable challenge for ICRP will clearly be that of integrating any approach to protection of the environment with that of the protection of human beings, bearing in mind that the latter is also the subject of a current, in-depth review. It is therefore of relevance that a number of different initiatives and concepts have been developing recently with respect to protection of the environment in relation to ionising radiation, both at national and international level. Much progress has been made in the last few years in the development of a variety of means for estimating exposures to a wide variety of animals and plants in different habitats. There has also been a high degree of co-operation amongst different researchers across many countries, encouraged by the IUR and financially supported in some cases by international bodies such as the European Commission (EC). A number of national programmes have also been significantly developed, and within at least one country, the USA, a legal basis has been established for applying dose-limit values in relation to certain nuclear sites. There is, therefore, already much being done but, although such programmes have many similarities, they also have the potential to diverge considerably and ultimately to be based on different principles, approaches, and scientific interpretation. Nevertheless, a common feature of many of these is, again, the concept of 'reference' models and data sets.

The Commission is therefore recommended to develop a small set of reference fauna and flora, plus their relevant databases—in a manner similar to that of Reference Man—

to serve as a basis for the more fundamental understanding and interpretation of the relationships between exposure and dose and between dose and certain categories of effect for a few, clearly defined types of animals and plants. It would also be useful if the magnitude of doses relating to these effects could be set out in a 'banded' fashion, such as the proposed 'derived consideration levels', in a manner similar to that being considered for human beings. Such a set of information could then serve as a basis from which national bodies could develop, as necessary, more applied and specific numerical approaches to the assessment and management of risks to non-human species as national needs and situations arise.

In this respect, it is also recognised that such assessment and management approaches differ from one situation to another, and each may constitute only a part of larger, existing environmental management programmes. Assessments may, therefore, be conducted for many different reasons, and situations may be managed in many different ways. Both will necessarily be integrated into other aspects of planning and action that may be expected to differ from one country to another. In many cases, such actions are already framed or constrained by existing legislation.

In conclusion, therefore, it is considered that ICRP can and should play the key role with respect to ionising radiation, both in advising on a common international approach, and in providing the basic interpretation of existing scientific knowledge in order for such a common approach to be delivered. The Commission is therefore urged to show its commitment to the protection of the environment, by way of protection of non-human species, and for this to be reflected in changes to its structure and work programme at the earliest opportunity.

1. INTRODUCTION

1.1. Aims

(1) The aims of this report are to address the specific requests of the ICRP by:

- defining how the ICRP can contribute to the attainment of society's goals of environmental protection by developing a protection policy based on scientific and ethical-philosophical principles;
- suggesting a framework for the assessment of the impact of ionising radiation in the environment based on science to support requirements for protection of the environment against harmful effects of ionising radiation; and
- showing how such a proposal for assessment of impact of ionising radiation in non-human species can be interfaced with or integrated into an overall system of radiological protection.

The report has been produced in the light of the ICRP's latest comprehensive set of recommendations for radiological protection found in its *Publications 60, 77, 81,* and *82* (ICRP, 1991, 1998a,b, 1999).

1.2. Scope

1.2.1. Extent

(2) The ICRP has not dealt explicitly with environmental protection before. Exposures of non-human organisms to radionuclides have been considered only in so far as they affect the radiological protection of humans. Hence, there are no ICRP recommendations as to why or how explicit protection of the environment with respect to radiation should be carried out, or what dose limits—if any—should be applied to other organisms. The ICRP is currently reviewing its existing recommendations for the protection of humans with the aim of developing its recommendations for the 21st century. This report provides a conceptual framework for the assessment of radiological impact of the environment that could feed into the next set of recommendations and support society's efforts to protect the environment.

(3) Environmental protection is influenced by a spectrum of cultural, ethical, and philosophical principles and views, and there has been substantial progress in this area since the preparation of *Publication 60* (ICRP, 1991). The increasing public concern over environmental hazards has led to the emergence of a variety of national and international legal commitments for protection of the environment. These commitments demonstrate a generally held view that an explicit means of demonstrating protection of biota and ecosystems from harmful effects of ionising radiation is also needed, and may often be legally required (e.g. Copplestone et al., 2001; Holm et al., 2002).

(4) The environment is composed of biotic and abiotic components that together form a system. Humans are part of this system and interact with both its living and

non-living components. Radiation interaction with living tissue is the most important component at ambient environmental dose rates. In this report, the focus is on living organisms (the biotic components of the environment) and the impact of ionising radiation upon them. Societal values of pollution prevention, precautionary principles, etc., are discussed to provide a context to the Task Group's recommendations, but the means of achieving these goals is beyond the scope of this report.

(5) This report is directed to the ICRP itself, and addresses the role that the ICRP could play in this important, developing area, building on the approach that has been developed for human protection and on the specific areas of expertise available to the ICRP, namely radiological protection. This approach aims to be free of national or political interference. The report does not, therefore, address what steps or measures could be implemented at national level, or how any particular industry or environmental circumstance should be managed or regulated. Instead, it examines and suggests what could be done by the ICRP—given our present state of knowledge—to provide an underpinning set of concepts, and reference methodologies, models, and databases, that could serve to provide a common basis for developing more detailed approaches to addressing the many issues that do, and will, arise with regard to the assessment of radiation impact on non-human species.

(6) The Task Group has concluded that a systematic approach for radiological assessment of non-human species is needed in order to provide the scientific basis to support the management of radiation effects in the environment. The Task Group recommends that the ICRP develops a framework for radiological assessment of non-human species that is harmonised with the proposed approach for the protection of humans. To achieve this, an agreed set of quantities and units, a set of reference dose models, reference dose-per-unit-intake data, and reference organisms will be required. The Task Group recommends that the ICRP, as a first step, develops a limited number of reference fauna and flora, so that others can develop more area- and situation-specific approaches to assess and manage risks to non-human species.

(7) The Task Group's report does not intend to define dose limits for biota, nor give recommendations on what to protect or on the level of risk that is acceptable. The proposed system does not intend to set regulatory standards. The Task Group rather recommends a framework that can be a practical tool to provide high-level advice and guidance in prospective situations, and help regulators and operators demonstrate compliance with existing legislation. The ICRP will address justification and optimisation in its recommendations for the start of the 21st century, and will then have to decide how protection of other species will affect these concepts.

1.2.2. Content

(8) Following this introduction, Chapter 2 gives a general overview of how environmental risks are perceived and addressed by society, and refers to the current theories of environmental ethics and some of the more important principles guiding protection of the environment generally. Chapter 2 also gives a brief summary of

how radiological protection of the environment is currently being developed by international organisations and national authorities. Chapter 3 gives a brief introduction to our understanding of radiation and living organisms, and Chapter 4 explores how the ICRP's current statement set out in paragraph 16 of *Publication 60* (*'The Commission believes that the standards of environmental control needed to protect man to the degree currently thought desirable will ensure that other species are not put at risk.'*) relates to modern environmental concerns. Chapter 4 also gives a number of reasons why the ICRP now needs to consider the basis upon which it can state more clearly its position and future role with regard to protection of part of the environment, namely, non-human species. In Chapter 5, various approaches to assess risks to the environment from hazardous effects of radiation are presented. The proposed system is designed so that it can be integrated with the approach being taken to protect humans and with those methods that are already in use or under development in some countries. These aspects are discussed in Chapter 6, including an indication of where further work is needed. Chapter 7 provides some concluding remarks.

1.3. Background

(9) In the 1970s, the ICRP believed that the major concern about radiation protection of species other than humans was to protect them simply as species or large populations rather than as individuals. As the policy of radiation protection of humans is to keep the risk to individuals very low, the ICRP concluded that this would probably keep radiation levels in general so low that other species living in the same environment would always be protected as species, if not as individuals. This conclusion was formulated in paragraph 14 of *Publication 26* (ICRP, 1977): *'Although the principal objective of radiation protection is the achievement and maintenance of appropriate safe conditions for activities involving human exposure, the level of safety required for the protection of all human individuals is thought likely to be adequate to protect other species, although not necessarily individual members of those species. The Commission therefore believes that if man is adequately protected then other living things are also likely to be sufficiently protected.'*

(10) Over the years, the ICRP has produced a large number of publications dealing with various aspects of the radiological protection of humans. The basic principles in the ICRP's current recommendations do not directly address protection of the environment. The present position of the ICRP is set out in paragraph 16 of *Publication 60* (ICRP, 1991): *'The Commission believes that the standards of environmental control needed to protect man to the degree currently thought desirable will ensure that other species are not put at risk. Occasionally, individual members of non-human species might be harmed, but not to the extent of endangering whole species or creating imbalance between species. At the present time, the Commission concerns itself with mankind's environment only with regard to the transfer of radionuclides through the environment, since this directly affects the radiological protection of man.'*

(11) In more explicit terms, the policy can be stated as follows.

- The ICRP's system of protection provides protection for humans. The system is not confined to dose limits.
- The application of the system of protection may sometimes damage or kill individual members of non-human species. The ICRP's policy has been to acknowledge this limited consequence.
- Although ecological information is incomplete, the full application of the system of protection is not thought to endanger whole species or to create imbalance between species. If this were not so, the ICRP's policy would be to require additional restrictions.

(12) This approach was not clearly set out and has been misinterpreted to mean that the ICRP's dose limits alone would be sufficient to protect non-human species. The ICRP has not claimed that the dose limits would be sufficient for this purpose. It also follows that the ICRP has not dealt explicitly with radiological protection of the environment, although non-human organisms may well have been afforded an indirect measure of protection as a result of the controls on radionuclide concentrations in environmental media established as part of the system of radiological protection of humans. Although methods and approaches are already available or being developed by some individual countries, there are no ICRP recommendations on appropriate assessment philosophies, methodologies, or guidelines on how radiological protection of the environment should be carried out. In particular, the ICRP has not advised on whether justification or optimisation should be considered in the cases of protection of species other than humans, or what dose limits—if any, and under what circumstances—should or could be applied to other organisms.

(13) Society's concern for environmental risks has put pressures on policy makers and regulators to define protection strategies that specifically and explicitly include the environment, as evidenced by a growing number of international and national legal commitments. This reflects both a need to protect the environment so as to maintain a suitable environment in which humans can exist, and a concern for the environment per se. In turn, these concerns reflect worries related to the possible effects of ionising radiation on the environment, as well as a desire to protect the environment simultaneously from a wide range of harmful influences. To meet the broader concern, strategies for protection of the environment are increasingly required to be applicable to radiation as well as to other pollutants.

1.3.1. The role of international organisations

(14) Many international bodies are involved in the radiological protection of humans, and in the case of environmental protection, even more organisations are of relevance. It is therefore useful to start with a short description of how these organisations work and relate to each other.

(15) The United Nations Scientific Committee on the Effects of Atomic Radiations (UNSCEAR) is the body within the United Nations (UN) system with a

mandate from the General Assembly to assess and report levels and health effects of exposure to ionising radiation. UNSCEAR was established by the UN General Assembly in 1955. It is composed of representatives from 21 nations and regularly publishes comprehensive reports on the levels and health effects of radiation. Governments and organisations throughout the world rely on UNSCEAR estimates as the scientific basis for evaluating radiation risk to humans, establishing radiation protection and safety standards, and regulating radiation sources.

(16) The ICRP is an independent registered charity that was established in 1928 by the International Congress of Radiology. The International Society of Radiology was formerly its parent organisation, although the ICRP's field of work has widened from protection in radiology to all aspects of protection against ionising radiation. The ICRP issues recommendations on the principles of radiological protection, and its recommendations form the basis for more detailed codes and regulations issued by other international organisations and by regional and national authorities.

(17) The International Commission on Radiation Units and Measurements (ICRU) was established in 1925 by the International Congress of Radiology. Its principal objective is to develop internationally acceptable recommendations regarding quantities and units of radiation and radioactivity, procedures suitable for the measurement and application of these quantities, and physical data needed in the application of these procedures.

(18) The International Atomic Energy Agency (IAEA) is an independent intergovernmental, science- and technology-based organisation within the UN. It serves as an intergovernmental forum for scientific and technical co-operation in the nuclear field, and as the international inspectorate for the application of nuclear safeguards and verification measures covering civilian nuclear programmes. The IAEA assists its member states in planning for and using nuclear science and technology for various peaceful purposes. It also develops nuclear safety standards and promotes the achievement and maintenance of high levels of safety in applications of nuclear technologies, as well as the protection of human health and the environment against ionising radiation.

(19) The Nuclear Energy Agency (NEA) is a specialised agency within the Organisation for Economic Co-operation and Development (OECD), an intergovernmental organisation of 27 industrialised countries. The NEA assists its member countries in maintaining and developing the scientific, technological, and legal bases required for the safe, environmentally friendly, and economical use of nuclear energy for peaceful purposes.

(20) The European Commission (EC) is the European Union's (EU) executive body, managing policies and negotiating international trade and co-operation agreements. It initiates EC policy and represents the general interests of the EU, acting as the guardian of the EU Treaties (which address inter alia environmental and radiological protection issues) to ensure that European legislation is applied correctly. Although the EC has a role as source of policy initiatives, all the major decisions on policies, actions, and legislation are taken by the ministers of the Member States in the Council of the EU, in co-decision (or, in some cases, in particular for legislation under the Euratom Treaty, consultation) with the European

Parliament. The EC's role is to ensure that the EU can work towards an ever-closer union of its members. There are many activities related to radiological and environmental protection. Several EC directives regulate radiological protection, and they are implemented through the national legislation of Member States. The EC also supports radiation research to be conducted in and amongst its Member States.

(21) The International Radiation Protection Association (IRPA) is an international organisation whose primary purpose is to provide a medium whereby those engaged in radiological protection may communicate and advance radiological protection. This includes relevant aspects of such branches of knowledge as science, medicine, engineering, technology, and law, to provide for the protection of humans and the environment from the hazards caused by radiation, and thereby to facilitate the safe use of medical, scientific, and industrial radiological practices for the benefit of mankind. A major task for the IRPA is to provide for and support international meetings for the discussion of radiological protection. The International Congresses of the IRPA itself are the most important of these meetings, which have been held about every 4 years since 1966. The IRPA also encourages the establishment and continuous review of universally acceptable radiological protection standards or recommendations through the international bodies concerned.

(22) The International Union of Radioecology (IUR) is a non-governmental organisation dedicated to making evaluations of the results and achievements in radioecology, and communicating these achievements to a broader audience.

(23) All of these international organisations relate to one another in different ways (Fig. 1.1). In short, UNSCEAR assesses studies published in the scientific literature. These assessments are then used by the ICRP as a basis for its own recommendations regarding human radiological protection. The ICRP also exchanges information and views with the ICRU and the IAEA. The IAEA interprets and converts the ICRP's recommendations into safety standards and practical guidelines for radiological protection, in collaboration with other organisations, such as the International Labours Organisation (ILO), the World Health Organisation (WHO), the Pan-American Health Organisation (PAHO), and the Food Agricultural Organisation (FAO). At regional and national levels, the ICRP recommendations are usually used as a basis for the derivation of relevant radiation protection legislation.

(24) In the case of protection of the environment, additional international organisations, e.g. the United Nations Environment Programme (UNEP) and the International Union for Conservation of Nature and Natural Resources (IUCN), play a role in defining the scientific, ethical, and legal bases for a system of protection and of the need for any standards.

1.3.2. Radiation and the environment

(25) At present, there are no internationally agreed criteria or policies that explicitly address protection of the environment from ionising radiation, although many international agreements and statutes call for protection against pollution generally, including radiation. Several international conventions and policies, as well as national policy statements and regulations, also demonstrate the need to develop

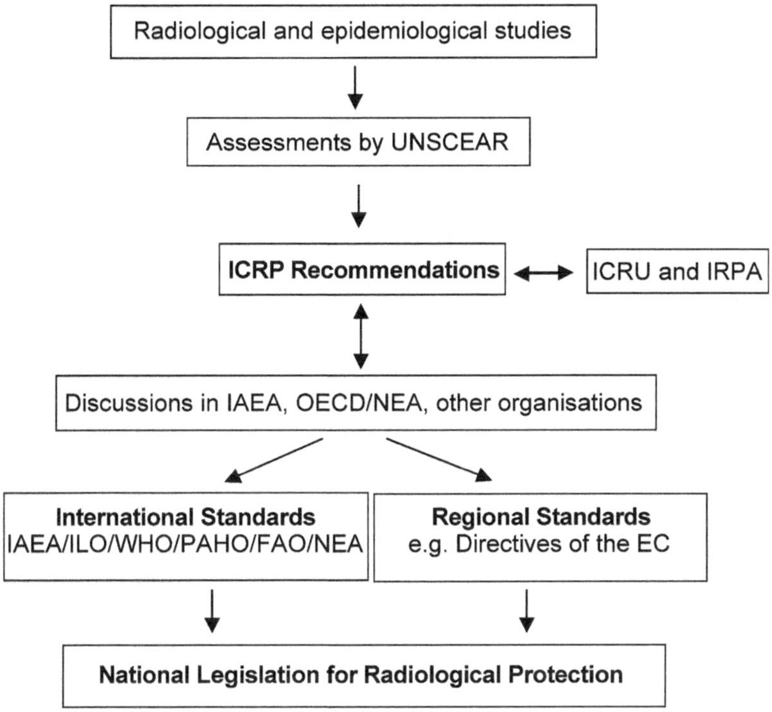

Fig. 1.1. From science to regulations on radiological protection of humans.

specific methodologies and criteria to support environmental protection decisions. Thus, although the existing ICRP system for protection of humans indirectly affords some level of protection to the populations of other species, the current approach lacks transparency, and the distributions of released radionuclides will always be such that other living organisms will receive dose rates different from those received by people. The lack of both a policy and a technical basis for assessment, criteria, or standards that have been endorsed at an international level makes it very difficult to determine or demonstrate whether or not the environment is adequately protected from potential impacts of radiation under different circumstances. The international development of an explicit assessment framework would support and provide transparency to the decision-making process. As mentioned earlier, this subject is currently being pursued in many countries with a view to developing or refining approaches for radiological assessment of the environment. A number of existing initiatives and developments in international organisations and agencies, as well as by national regulatory bodies, therefore already provide an input as to how one might best proceed.

(26) Several international conventions emphasise the need for protection of the environment with respect to radiation. The Joint Convention on the Safety of Spent Nuclear Fuel Management and on the Safety of Radioactive Waste Management (Waste Convention, 1997) was set up with the co-operation of the IAEA. The

Convention was adopted in 1997 and came into force in June 2001. It aims at protecting individuals, society, and the environment against the harmful effects of radiation, and includes the following statement: *'Each Contracting Party shall take appropriate steps to ensure that at all stages of spent fuel management (radioactive waste management), individuals, society and the environment are adequately protected against radiological hazards.'*

(27) As part of the OSPAR Convention for the Protection of the Marine Environment of the North-East Atlantic (OSPAR, 1992), Contracting Parties agreed to *'take all possible steps to prevent and eliminate pollution and to take necessary steps to protect the maritime area against adverse effects of human activities so as to safeguard human health and to conserve maritime ecosystems...'*. Further, the OSPAR Strategy with Regard to Radioactive Substances, decided at the Ministerial Meeting of the OSPAR Commission in Sintra in 1998 (OSPAR, 1998), agreed to the objective to *'prevent pollution of the maritime area from ionising radiation through progressive and substantial reductions of discharges, emissions, and losses of radioactive substances with the ultimate aim of concentrations in the environment near background levels for naturally occurring radioactive substances and close to zero for artificial radioactive substances'*. In practice, the Strategy means that by the year 2020, discharges and emission of radioactive substances should be reduced to levels where the additional concentrations in the marine environment above historic levels, resulting from such discharges, emissions, and losses, are close to zero. The Strategy also requests the OSPAR Commission to undertake the development of environmental quality criteria for the protection of the marine environment from the adverse effects of radioactive substances, and report on progress by the year 2003.

(28) The scope of the current *International Basic Safety Standards for Protection against Ionizing Radiation and for the Safety of Radiation Sources* (IAEA, 1996) is limited to protection of human beings, and following the line in ICRP *Publication 60*, it states that: '*it is considered that standards of protection that are adequate for this purpose will also ensure that no other species is threatened as a population, even if individuals of the species may be harmed.*' However, the IAEA Safety Fundamentals *The Principles of Radioactive Waste Management* (IAEA, 1995) includes the requirement that '*radioactive waste shall be managed in such a way as to provide an acceptable level of protection of the environment*'. Other principles reflect the concern for sustainability and the right of all other states to be consulted.

(29) The IAEA has also addressed the explicit issue of environmental protection on several other occasions. In 1970, a panel of experts was convened by the IAEA to assess the principles for limiting the introduction of radioactive waste into the sea. Among other things, this panel recommended pursuing '*the study of the effects of ionising radiation on organisms and their sensitive life stages with special regard to effects at the genetic, population and ecosystem level*'. Subsequently, several expert meetings were held on the subject resulting in the publication of the report *Effects of Ionizing Radiation on Aquatic Organisms and Ecosystems* (IAEA, 1976). Other IAEA work, in support of the London Convention in 1972, explored the possible effects of sea dumping of radioactive waste packages on marine species and, in 1979, the IAEA published the report *A Methodology for Assessing Impacts of Radioactivity on*

Aquatic Ecosystems (IAEA, 1979). A further report, *Assessing the Impact of Deep Sea Disposal of Low Level Radioactive Waste on Living Marine Resources* (IAEA, 1988), discussed the doses to a number of 'typical' marine species living at or near the sea floor. In 1992, a report, *Effects of Ionising Radiation on Plants and Animals at Levels Implied by Current Radiation Protection Standards*, dealing with the effect of radionuclide releases on terrestrial and freshwater environments was published (IAEA, 1992).

(30) In 1999, the IAEA published the report *Protection of the Environment from the Effects of Ionizing Radiation* (IAEA, 1999) that presented various issues and approaches for establishing an environmental protection framework and criteria. More recently, the report *Ethical Considerations in Protecting the Environment from the Effects of Ionizing Radiation* was published (IAEA, 2002a). The IAEA is continuing work towards the development of a safety standards document on environmental radiation protection, in collaboration with other international organisations. It also continues to foster information exchange by holding specialist meetings on the subject, the most recent of which took place in November 2001 (IAEA, 2002b).

(31) At that meeting, the participants agreed that '*it is necessary to develop a system for the radiological protection of the environment (or the biotic components of it)*'. The specialist meeting saw a need '*to distinguish protection of biota from protection of the environment, which includes abiotic components. However, it was agreed that the initial focus should be on the protection of biota*', and recognised the need for international co-ordination and co-operation (IAEA, 2002b). In its report to the Director General of the IAEA, the meeting encouraged the IAEA '*to continue working towards the development of safety standards that are practically base*', and identified the IAEA as having '*a potentially valuable role in the consideration of the way in which effects manifested in individuals are expressed on higher levels of organisation (populations, communities and ecosystems), and in the development of a compilation of transfer factors from different sources*'. The meeting also agreed that '*the use of reference organisms is a reasonable approach to adopt in the development of a system to protect biota from the effects of radiation*', and recognised that '*effects on higher levels of organisation (e.g. populations) occur only if individual organisms are affected, and that effects data are generally available for individuals rather than higher levels of organisation*' (IAEA, 2002b).

(32) In 1996, UNSCEAR published a comprehensive report on the effects of radiation on the environment, taking into consideration the specific problems encountered with dosimetry and quality factors for non-human biota, experience from experimental studies, observations made in certain environments as a result of routine discharges, as well as observations made after accidental releases (UNSCEAR, 1996). The report summarised a large amount of work that had been done on this subject for many decades, and serves as a scientific background document to the development of standards and recommendations by regulatory bodies.

(33) The OECD/NEA have summarised environmental and ethical principles of geological final waste disposal (OECD, 1995), and have recently identified the need to clarify the ICRP's current view on environmental protection (OECD, 2000). The

NEA is organising three international fora in collaboration with the ICRP to discuss radiological protection of the environment. The first forum took place in February 2002 with the objective of developing—together with participants from 20 countries and seven international organisations, representing regulators, politicians, science, industry, and international organisations such as the IAEA, EC, WHO, ILO, and non-governmental organisations such as the IRPA and Greenpeace International—a sound technical basis and criteria for ICRP recommendations on radiological protection of the environment (OECD, 2002). These meetings will involve a wide range of stakeholder views, and will help to ensure that the Commission's recommendations for protection of the environment will provide benefit to the environment whilst also being balanced against the benefits to society in an overall practical system of protection.

(34) In 1997, the IUR undertook a concerted action for the EC. The initial results were published in 2000 with the conclusion that a framework for protection of non-human species was urgently required in order to structure the knowledge derived from earlier studies (Pihet, 1998; Strand et al., 2000). A preliminary approach for environmental radiation protection was identified in order to direct future scientific research, which included the derivation and development of relevant quantities and units, reference organisms, environmental transfer models, reference dosimetric models, and tabulated dose rate/effects information for reference organisms. The IUR is collecting information on research activities and priorities for future work, and was one of the organisers of a 'Consensus Conference on Protection of the Environment' in October 2001. A consensus statement from that conference (IUR, 2001) included the following guiding principles: *'Humans are an integral part of the environment, and whilst it can be argued that it is ethically justified to regard human dignity and needs as privileged, it is also necessary to provide adequate protection of the environment. In addition to science, policy making for environmental protection must include social, philosophical, ethical (including the fair distribution of harms/benefits), political and economic considerations. The development of such policy should be conducted in an open, transparent and participatory manner. The same general principles for protection of the environment should apply to all contaminants.'*

(35) In 1997, the Arctic Council (consisting of the Nordic countries, Russian Federation, Canada, and the USA) identified the need for developing an assessment and protection framework for the protection of the environment in the Arctic. This has led to a programme during the period 1998–2002 and a report was approved at the Ministerial Conference of the Arctic Council in October 2002 in Finland. The Arctic Council has co-operated with the IUR and the EC on this issue, and has also endorsed the initial focus on protection of the biota and the use of reference organisms. The issue of environmental protection (with an initial focus on protection of biota) was also addressed at a Ministerial Conference on the North Sea in April 2002. The conference endorsed the ongoing work of the IUR and others to achieve an international consensus for a framework for the protection of the environment.

(36) There are a number of EC directives that relate to radiological and environmental protection, although the basic safety standards for the public and workers (the 96/29/EURATOM directive, CEC, 1996a) focus on doses to, and protection of,

humans. Examples of European directives of relevance for environmental protection generally are the Directive on Integrated Pollution Prevention and Control (CEC, 1996b), the Directive of the Conservation of Natural Habitats and of Wild Fauna and Flora (CEC, 1992), the Water Framework Directive (CEC, 2000), and Directive 85/337/EEC on Impact of Certain Projects on the Environment (CEC, 1985). The latter is designed to ensure that, before consent for the development of a project is given, projects that are likely to have a significant effect on the environment (because of their nature, size, or location) are made subject to an assessment with regard to their expected effects. Environmental impact assessments must consider humans, fauna, flora, the abiotic environment (soil, water, air), material assets, and cultural heritage as well as the interactions amongst these factors. A study on the scope and application of 85/337/EEC, specifically in relation to geological disposal of radioactive waste, was presented at the IAEA's Conference on the Safety of Radioactive Waste Management, Córdoba, Spain, 2000 (Webster, 2000). By insisting on an environmental impact assessment for substantial projects, 'best practice' is demonstrated and enables consideration of the benefits of harmonisation of approaches in different countries.

(37) In view of the increasing awareness in the EU of the need for a system to demonstrate protection of the environment and current work on demonstration of protection of biota, the EC is funding scientific research in this area (Strand and Larsson, 2001). For example, the Framework for Assessment of Environment Impact (FASSET) programme aims at obtaining a scientific basis for judging the likelihood or not of radiation damage to biota in the context of protecting humans and the environment. A study of the Environmental Protection from Ionising Contaminants in the Arctic (EPIC) is also underway, again funded by the EC, looking at environmental radionuclide transfer in the Arctic, modelling uptake by biota, identifying reference biota to evaluate potential doses and dose-effect relationships, and integrating assessments of environmental impact with those for other contaminants.

(38) Much has already been learned from the work of national programmes, particularly in the USA, Canada, Russia, the UK, and France. But it is only in one country, the USA, where an authority—the US Department of Energy (USDOE)—has developed requirements and guidance for the radiological protection of the environment, has currently in place a radiation dose limit for protection of aquatic biota, and has proposed limits for protection of terrestrial biota (USDOE, 1996) for some of its own facilities. The USDOE developed screening methods using a set of reference organisms within a graded approach for demonstrating protection of biota applicable to these dose-rate guidelines. In addition, the results from the current EC multinational research projects FASSET and EPIC are expected in 2003, and will allow the production of systematic frameworks—using a 'reference fauna and flora approach'—to assess environmental impact in different ecosystems. In Canada, the Canadian Nuclear Safety Commission (CNSC) is proceeding in a similar direction with guidance being developed on an integrated framework to assess the impacts of ionising radiation and other environmental contaminants to non-human organisms (e.g. Bird et al., 2003). These frameworks are intended to provide a scientifically based assessment approach to inform decision makers and stakeholders. It is

anticipated that the results of these programmes will, together with the results of other national and international work, contribute towards the development of international recommendations and guidance. The development of a broad international consensus and the inclusion of recommendations and guidance in national legal instruments is therefore likely to continue beyond 2003.

(39) In summary, therefore, it would appear that there has clearly been a shift in society from the long-held anthropocentric approach to matters environmental to one that embraces both biotic and abiotic components of the environment. All of the recent conventions, principles, reports, and statements lend support to the now widely held view that there is a need to demonstrate, explicitly, that the environment can and will be protected from the effects of radiation.

2. CURRENT ENVIRONMENTAL MANAGEMENT PRINCIPLES

2.1. How are environmental risks currently addressed by society?

(40) Environmental risks, their identification, and their management are all part of modern life. The environment of which we are part is largely a managed one, both with respect to what we remove from it and what we put into it as a result of virtually all of our daily activities. But such is the scale of this human impact that many measures and steps are taken to 'protect' the environment in one way or another. Such steps are not necessarily co-ordinated, and thus it is reasonable to ask: what does *environmental protection* actually mean? It is also not simply a scientific question, because it can only be answered by reference to both qualitative and quantitative evaluations of the environmental effects against which protection is being afforded, plus an evaluation of whether or not this matters, and to whom. A convenient starting point, therefore, is to examine the possible ethical basis—or bases—of environmental protection, and its links with scientific and legal aspects of the subject. An Advisory Group has recently undertaken such work for the IAEA (IAEA, 2002a), testing its findings though a series of specialist meetings with participants from IAEA Member States and from various international organisations, including the ICRP. Its conclusions are, briefly, as follows.

(41) Ethical considerations are clearly important in the derivation of concepts such as environmental protection. Even for the protection of humans, the need for which is generally accepted without question, different ethical considerations have had, and will continue to have, an important part to play. For example, in providing a system of protection with respect to ionising radiation, the ICRP's 'ALARA' basis for optimising the level of protection has been seen as being consistent with, and a consequence of, a *utilitarian* ethic, i.e. the greatest good for the greatest number, whereas its constraint by the application of a dose limit has been seen as consistent with a *deontological* ethic, i.e the rights of, and duty towards, individuals (Shrader-Frechette, 1994; ICRP, 2001b).

(42) Different ethical views have similarly affected the way in which people view the environment, their impacts upon it, and how best to manage the consequences. Such different ethical views have resulted, in turn, in the emergence of different social, cultural, religious, and legal differences across the world (Rawls, 1971; De Shalit, 2000). Thus any systematic approach to addressing the issue of how best to protect the environment with respect to ionising radiation has to accommodate such views, and their consequences, as best it can.

(43) In the IAEA study (2002b), a useful three-component ethical spectrum of views was identified: *anthropocentric*, *biocentric*, and *ecocentric*. These views arise from philosophical debates about what has moral standing in the world and why. Essentially, and grossly oversimplifying the subject, these three views may be summarised as follows:

- ***anthropocentric***, in which human beings are the main or only thing of moral standing, and thus the environment is of concern only as it affects humans;

- ***biocentric***, in which moral standing can be, and is, extended to individual members of other species, and thus obligations pertaining to such individuals arise as a consequence; and
- ***ecocentric***, in which moral standing can be extended to virtually everything in the environment, including landscapes—rivers and mountains—but the focus lies more with the entirety and diversity of the ecosystem rather than, say, the moral significance of each and every individual component of it.

There are, of course, considerable ranges of views within each of these three broad categories.

(44) The anthropocentric view is the most easily recognised; the other two less so. Biocentric views vary considerably, but a common feature of many of them is recognition of the moral obligations that arise from the fact that, for example, many animal species can be shown, 'scientifically', to be *sentient*, in that they can experience pleasure and pain. The results of these considerations are reflected in attitudes to animal 'rights' and animal 'welfare', and thus in national laws—such as those relating to experiments on animals, for whatever reason. Biological characteristics other than sentience may also be considered relevant, and some biocentric views assume that all individual living things have an inherent value and should be respected for what they are. Those with an ecocentric view believe that one should optimise ecosystem welfare, and although they may disagree about how to carry out such an optimisation, they agree that primacy, in moral standing, rests with ecosystems. The place of humans and the degree to which they can be considered to have special 'rights' compared with those afforded to other species and to physical components of the environment also vary. Such views can often be clearly recognised in many cultures and beliefs. It also has to be admitted that individuals may change their ethical views during their life, or when faced with different circumstances. But such views are also—and importantly—collectively reflected at social, cultural, and religious levels of society.

(45) Nevertheless, in view of such basic ethical differences and attitudes to the environment, it is also reasonable to ask: can one identify any common ground for a consensus on such issues? The IAEA Advisory Group addressed this question by examining the nature and content of multilateral environmental agreements that have emerged in recent years, the signatories of which not only represent different cultures from all over the world, but indicate how these are reflected—at a national level—in their attitudes to matters environmental (IAEA, 2002b). The following areas of agreement were considered to be particularly relevant.

- ***Sustainable development***. The UN 'Rio' Declaration of 1992 brought this concept into prominence (UN, 1992). Sustainable development relates to the need to recognise the interdependence of economic development, environmental protection, and social equity, and thus the obligation also to protect and provide for both the human and environmental needs of present and future generations. It includes, and explicitly refers to, a number of other concepts, including the precautionary principle.

- ***Conservation***. There are many international agreements relating to the conservation of both species and habitats. They essentially relate to the 'importance' or 'vulnerability' attached to individual species, or areas where many species live, particularly with regard to the need for agreement at an international level in order to protect them; for example, the need to ensure that migratory species can safely travel and survive throughout their natural migratory range.
- ***Preservation.*** Preservation recognises the worth of nature as pristine, as independent of human needs. Preservationists also argue for the value of wilderness, land untouched by human degradation or resource use; they recognise that wilderness is an important cultural value, not only in itself but also with respect to promoting character, spirituality, and natural systems (NRC, 1993). In many countries and internationally (e.g. UNESCO), the principle of preservation has led to the establishment of natural preservation areas in which human activities are strictly controlled.
- ***Maintenance of biodiversity***. This obligation also stems from Rio (UN, 1992), and recognises the need to maintain the biological diversity inherent within each species, amongst different species, and amongst different types of habitats and ecosystems.
- ***Environmental justice***. Another feature of the Rio Declaration is the explicit responsibility to ensure that activities within national jurisdiction or control do not cause damage to the environment of other states. This, in turn, reflects the general principle of environmental justice: the need to take account of the fact that inequity can and does arise between the distribution of what might be termed 'environmental benefits and harm'. Where such differences amongst nations occur, it is expected that they be addressed either by re-distributing the benefits, or by compensating for the harm. Such actions are, admittedly, more about how one goes about achieving environmental protection than defining what it actually is. But the concepts behind them are very important. The imbalance of benefits and harm across national borders (such as transboundary pollution) is relevant to the concept of *distributive* justice (or injustice), and the need for restitution or compensation for such pollution is relevant to the concept of *retributive* justice. It is also relevant to note that inherent in both concepts is the implicit ability to quantify damage to the environment, plus the moral need to restore it, or to compensate in some other way, when it has been damaged.
- ***Human dignity***. This, too, is a concept upon which there is international agreement. It is the cornerstone of the Charter of the UN (UN, 1945). It also has relevance to the concept of environmental protection and how it can be achieved. It recognises the need for the respect of individual human rights, and for the consequent range of human views. It therefore axiomatically requires an anthropocentric view to recognise the existence of, and the equal validity of, both biocentric and ecocentric views within societies, and thus the obligation of taking them into account in a process of informed consent. It also recognises that there will be different personal views—irrespective of the existing legal position—about the way protection is currently afforded to

other living things, either as individuals or as part of an ecosystem, and why. In addition, it recognises that the dignity of people may be challenged and offended because of the disturbance of the environment in all manner of different ways — such as the *presence* of 'unnatural' chemicals in the natural environment, irrespective of any known effect they may have on the living components of it, or even in the absence of any living component.

(46) As one might imagine, there is no simple way of distilling all of these 'principles' into a single ethic for environmental protection. But the above set areas help to frame the concept of what society currently means by environmental protection, and how it might be achieved, within which more specific aspects could usefully be addressed.

(47) Many methodologies and regulations to protect the environment have been developed over many years, notwithstanding the fact that our understanding of ecology is incomplete, as is our understanding of the impact of environmental pollutants generally. Consideration of these limitations has resulted in the adoption of several 'principles' with the purpose of protecting the environment. The most relevant of these operational strategies can be summarised as follows, although there are variants in their formulation and practical application.

- ***The pollution prevention principle***, which argues that emissions should be controlled to the extent practical, taking socio-economic factors into account.
- ***The precautionary principle***, which argues that where there are threats of serious or irreversible damage, a lack of full scientific certainty should not be used as a reason for postponing cost-effective measures to prevent environmental degradation.
- ***The principle of using best-available techniques and technologies***, which argues that discharges into the environment should be kept to a minimum by employing the most robust techniques and managerial procedures that are available and economically feasible, even in situations where the benefit of such actions may be difficult to assess in terms of environmental harm or direct economic benefit (as may be the case in environmental protection).
- ***The substitution principle***, which argues that where safer alternatives are already available, or may be marketed in the near future, these should be promoted as a substitute to the activity/product in question. It therefore allows technology-driven changes (best environmental option) to improve environmental protection instead of waiting for the proof of harm.
- ***The polluter-pays principle***, which argues that 'polluters' are responsible for the environmental and economic effects of their 'polluting' activities. This principle was first widely discussed in the UN Conference on Environment and Development held in Rio in 1992. All the attending representatives of the countries and nations endorsed it.
- ***The principle of informed consent***, which emphasises the need for communication and public involvement, starting at the planning stage and well before decisions are taken from which there is no return. Such transparency of decision making should enable analysis and understanding of all stakeholders'

arguments, although decisions against certain stakeholders may not be avoided. Transparency is usually secured by way of an environmental impact assessment.

(48) The various applications of these principles have resulted in regulations for environmental protection that combine minimisation of environmental effects based on scientific evidence and pollution prevention to the extent that is achievable based on social and economic considerations. In good environmental practice, the goal should be a clear separation between science (i.e. the assessment of environmental effects, including the consideration of uncertainty and variability) and management phases (socio-economic factors determining the implementation of mitigation measures and the setting of environmental management objectives) of environmental protection, although in reality, the scientific and management aspects cannot be separated because of uncertainties and variations in the data, models, and frameworks (NRC, 1996).

(49) Regulatory requirements for protection of the environment are often written in terms of '*no significant adverse effect on the environment*', or that substances should not enter the environment in quantities, concentrations, or under conditions that have or may have an immediate or long-term '*harmful*' effect on the environment itself or its biological diversity. Environmental assessment methods (e.g. ecological risk assessment) must therefore be capable of demonstrating whether or not such environmental objectives will be met by the proposed control over a given industrial activity, and of describing the level of environmental harm when effects are predicted to occur. This has sometimes required the development of environmental protection benchmarks (e.g. limits, criteria, standards) that are representative of '*no expected effects*' on the environment against which predicted or observed environmental pressures can be compared. When actual or potential environmental values exceed these benchmarks, a quantification (with an indication of the level of uncertainty) of potential effects is needed.

(50) Any framework for environmental protection that is developed for radiation therefore needs to acknowledge and accommodate the principles outlined above, and needs to be compatible with other environmental protection approaches that will be in place for non-radiological emissions from these same facilities or other industrial practices.

2.1.1. Sustainable development

(51) The 1972 UN Conference (UN, 1972) on the Human Environment in Stockholm was the first international conference to lay down principles for the protection and improvement of the human environment. Then, in 1980, the World Conservation Strategy was published (IUCN, 1980), having been commissioned by the UNEP and prepared by the International Union for Conservation of Nature and Natural Resources (IUCN). Its aim was to help advance the achievement of sustainable development through the conservation of living resources, because it was recognised that this was essential to human survival, and thus to the concept of sustainable development. It also identified the priority conservation issues, plus the main requirements for dealing with them.

(52) The Brundtland Report (World Commission on Environment and Development, 1987) further alerted the world to the urgency of making progress towards a form of economic development that could be sustained without depleting natural resources or harming the environment. The concept of sustainable development was defined as '*development that meets the need of the present without compromising the ability of future generations to meet their own needs*'. By and large, this or similar definitions have been accepted in other international fora, as well as by national authorities. The report also emphasised the need to preserve biological diversity or 'biodiversity'.

(53) The 1992 UN Conference on Environment and Development in Rio (UN, 1992) then laid down a number of general principles for environmental protection, e.g. the Rio Declaration, the Convention on Biological Diversity, and the Work Programme Agenda 21. The Rio Declaration emphasised, amongst other things, that protection of the environment has to be an integral part of the sustainable development concept. The Convention on Biological Diversity (UN, 1992) similarly stressed the importance of recognising that all organisms contribute to the structure of the ecosystem. It defined the concept of biodiversity as *'the variability amongst living organisms and the ecological complexes of which they are a part, and thus the diversity within species, between species, and of ecosystems'*. Both the Declaration and the Agenda 21 Programme of Action called for governments to undertake national assessments of their biodiversity and formulate strategies to preserve and sustain it.

(54) Since 1992, the concept of sustainable development has increasingly affected the many practical considerations and decisions that continually have to be made at national level, including efforts aimed at protecting the environment. But the concept itself does not define the ultimate goal for development, because this may change according to changes in societal needs, and because opportunities for conservation and preservation may change. Limitations will therefore always be imposed on society's ability to take actions, on primarily economic and social grounds, although there is the important obligation to pass on a full range of present options to future generations. This emphasis on the dimension of time is essential for sustainable development. It also implies that practices cannot be analysed in isolation from a life-cycle perspective of their activities.

(55) Societal and economic factors may often clash. Thus the presence of low levels of contamination in environmental media may affect economic assets by reducing the market value of the products of affected regions, even though the detriment per se may be perceived to be within acceptable levels at any one time. Again, a robust framework for assessing and managing environmental risks, and for communicating these issues to the public, is a prerequisite to informed decision making. Sustainable development therefore necessitates that full consideration is taken of environmental factors, as well as of economic, ethical, and social factors.

2.1.2. Biological resources and biological diversity

(56) Protection of public health and protection of the environment can be regarded as complementary—or alternative—endpoints within the overall framework of sustainable development. However, while the protection of public health may have

well-defined endpoints, protection of the environment potentially involves a wide range of biological endpoints.

(57) Again, from the point of view of sustainable development, there are two aspects that need to be considered: the maintenance of biological diversity and the conservation of biological resources. Maintenance of biological diversity is one objective for all actions taken to protect the environment, whereas the conservation of biological resources is a necessity for human survival and general well being. But the two must be related. Thus Principle 4 of the Rio Declaration (UN, 1992) describes sustainability in this context: '*...the use of biological components of biological diversity in a way and at a rate that does not lead to the long-term decline in biological diversity, thereby maintaining its potential to meet the demands of present and future generations.*'

(58) The definition of biological diversity given in the Convention on Biological Diversity (UN, 1992) also emphasises its importance to the function of different ecosystems that, of course, also collectively constitute the environment within which humans live. Nevertheless, biological diversity is not static, but dynamic and continuously changing. Preservation of biological diversity thus does not mean conservation of a certain state, but protection against harmful effects that would cause diversity to develop in a fashion that would not otherwise have been the case. The UN Conference in Rio has also defined biological resources as '*the genetic resources, organisms or parts thereof, populations or any other component of ecosystems with actual or potential use for humanity*' (UN, 1992).

(59) The principles described above have supported the development of international conventions and national legislation that place requirements on the control of human activities, including the release of radionuclides to the environment. Given that any system of radiological protection of the environment must fit into the legal, ethical, and political framework set by society, the ICRP must ask itself what role it can play to support society's objectives. Obviously the ICRP cannot play the same role in environmental protection that it has played and is playing in the radiological protection of human beings, where its recommendations have founded the basis of national legislation. Consequently, the following chapters describe how the ICRP's specific expertise in radiobiology, dosimetry, etc. can best be used to support international efforts in the radiological protection of the environment.

2.2. Assessment and management of environmental risks

(60) In order to properly address society's demand for protection of the environment, general frameworks have been developed for assessing and managing environmental risks. The whole process of environmental risk assessment and management can be divided into three stages, here for convenience termed 'problem formulation', 'risk assessment', and 'risk management', although their application can vary at a national level (e.g. Barnthouse, 1995; Jones et al., 2002).

(61) The problem formulation stage involves scientific judgements related to the identification of sources and hazardous substances, and of possible interactive effects with other contaminants and certain ecological functions. It is clear that the

stage of problem formulation to a large, and sometimes dominating, extent is guided by societal views on what needs to be protected, and any assessment and management framework must be able to respond to society's demands in order to be credible and operational. Specific protective legislation may cover factors such as sustainable development, air, water, ecosystems, endangered species, and organisms of high cultural esteem or of economic value, which are bound to have a substantial impact on the outcome of the problem formulation.

(62) The risk assessment stage involves the execution of methods for analysing exposure and effects that are considered most suitable to the previously defined purpose. The ultimate result is the characterisation of the risk to which the subsequent management actions need to be tuned.

(63) The risk management stage involves any decision or action that will result in the prevention, mitigation, or elimination of environmental consequences, i.e. environmental protection. Several concepts and principles are either implicitly or explicitly taken into account in the development of modern approaches and regulations to protect the environment. In contrast to the ICRP's current protection philosophy, where consideration is given to radiation exposures that may result from a practice or an intervention, environmental protection approaches have evolved to take into account the fact that non-human biota may be exposed simultaneously to many stressors (radiological, chemical, thermal, etc.) potentially present in industrial effluents.

3. BIOLOGICAL EFFECTS OF RADIATION IN NON-HUMAN ORGANISMS

(64) The majority of our information on the exposure and effects of radiation relates to, and has been obtained to serve the needs of, the radiological protection of human beings. Similarly, much of our information on the behaviour, effects, and distribution of man-made radionuclides in the environment has also been derived to meet the needs of human radiological protection. There are, however, very considerable differences in the means by which species other than man may be exposed to radiation, even when all are present in the same environment at the same time. Hence there may be differences in the resultant doses and dose rates received amongst different plants and animals (by several orders of magnitude), and in the types of tissues and organs that receive the dose. Different types of radiation, from external and internal sources, will also result in different tissues and organs being exposed, all of which will result in different biological consequences for different types of animals and plants. This is particularly the case for alpha- and beta-emitting nuclides. Such variations have already been the subject of many reviews. This chapter therefore briefly considers some of the common features of the effects of radiation on living things.

(65) DNA is the critical primary target for the induction of biological effects of radiation in all living organisms. This has been evidenced from many radiobiological studies using various types of cells from animals and plants. The diameter of all DNA molecules is about 2 nm and, in terms of sensitive structures for energy depositions, this leads to broad similarities in radiation responses for different organisms. Thus the wealth of data that has been published about initial radiation mechanisms of relevance for humans (UNSCEAR, 1986, 1996, 2000) probably relates to many other organisms as well.

(66) Ionising radiation induces many different kinds of DNA damage, and not all of them will be equally important to the final cellular effects that are of interest for radiological protection. Lesions that may lead to loss or alterations of the genetic information in DNA during repair are considered to be the most critical radiation-induced DNA damage in terms of cell inactivation, mutation, chromosomal aberrations, and cell death. Of particular importance are chemically complex DNA double-strand breaks which are believed to be difficult to repair correctly (Goodhead, 1994; UNSCEAR, 2000).

(67) Although the cells of most mammals have roughly the same amount of DNA, cells show considerable differences in radiosensitivity. Radiosensitivity also varies with respect to the phases of the cell cycle, and cell-cycle regulation is an important contributor to cellular radiosensitivity. Indeed, a dogma in radiobiology, formulated as long ago as 1906 by Bergonié and Tribondeau, states that cells are radiosensitive if they are mitotically active, undergo many cell divisions, and are functionally undifferentiated.

(68) In mammals, most cell production occurs in the bone marrow and the small intestine. For other tissues, such as those of the central nervous system, radiosensitivity is greatest during early development when the neuroblasts are proliferating. It may be expected that the radiation response of these tissues in mammals,

and possibly all vertebrates, is similar to that of humans. Radiosensitive tissues for other organisms may be quite different. The radiosensitive parts of plants are usually the meristem tissues, which are located in the roots and shoot tips and, in trees, in an annulus around the trunk. This superficial location of the meristem makes it particularly vulnerable to radiation exposure from the deposition of radionuclides (UNSCEAR, 1996).

(69) The concept of absorbed dose gives a good description of the energy deposition in biological systems. At low doses or dose rates, there will be a heterogenous spatial distribution of the energy deposited, and radiation causing heterogeneous energy distributions will give different biological effects for the same adsorbed dose (Van der Stricht and Kirchmann, 2001). This difference may be quantified by applying a relative biological effectiveness (RBE) factor that relates to a defined biological endpoint in a specified organism or tissue.

(70) For humans, a radiation-weighting factor is derived from information about RBE, but the two quantities are very different. The radiation-weighting factor is a generalised quantity representing all the relevant endpoints in all the tissues of the body, and there is a considerable element of judgement involved in its derivation. The Commission's 1990 recommendations for humans recommend a radiation-weighting factor of 1 for photons and electrons, 5 for some neutrons (<10 keV and >20 MeV) and protons, and 20 for alpha particles (ICRP, 1991).

(71) There has recently been much interest in the need for similar concepts and values for animals, and indeed plants, particularly in relation to the expected relative effects of high-LET radiation. Many ranges of values have been suggested (e.g. UNSCEAR, 1996; Kocher and Trabalka, 2000; Trivedi and Gentner, 2000; Pentreath and Woodhead, 2001; ACRP, 2002; Thompson et al., 2003).

(72) High radiation doses may kill a large number of cells, thereby impairing the function of vital organs and tissues. *Deterministic* harm occurs above a certain threshold dose, and the severity of the effects increases with dose. Cancer or hereditary effects are *stochastic* effects, usually caused by damage in a single cell, and the probability of induction—but not the severity—is assumed to be proportional to the dose in the low-dose and low-dose-rate region. The stochastic effects are therefore assumed to have no threshold in humans (ICRP, 1991). There is increasing support from mechanistic studies for this assumption (UNSCEAR, 2000). For the purpose of protection of species other than mammals, it would probably be premature at this stage to try to distinguish between deterministic and stochastic effects. Radiation effects could therefore best be grouped into several broad categories, such as early mortality (the organism dying earlier than it otherwise would have done), 'morbidity' (a reduction in general physical and/or mental well being including effects on growth and behaviour), and reduced reproductive success (including effects on fertility and fecundity). Morbidity and reproductive disturbances are generally assumed to occur at much lower doses than mortality.

(73) Radiation may also cause damage that can be transmitted to subsequent generations. For humans, UNSCEAR has estimated the risk estimate of hereditary effects in the offspring of exposed individuals to be about 10% of the cancer risk of the exposed parents (UNSCEAR, 2001). For non-human organisms, it is even more

difficult to interpret the significance of hereditary effects at the population level (i.e. population fitness and survival) due to natural selection. Only if mutations confer a selective advantage in connection with a particular environmental condition will they spread in the population. 'Deleterious' mutations will generally be selected against in the population; 'neutral' mutations may persist over many generations. This concept of mutation—selection balance—has been discussed by UNSCEAR (2001).

(74) All of the above categories comprise many different radiation effects on individual organisms, and collectively they reflect the limitations of our current knowledge. They are, nevertheless, similar to the endpoints that are often used for risk assessments of other environmental stressors, and are relevant to the needs of nature conservation and other forms of environmental protection.

(75) Effects on higher levels of biological organisation (e.g. populations and ecosystems) occur only if individual organisms are affected, and effects data are generally obtained for individuals rather than for higher levels of organisation. Caution should be made for situations where the effects on individuals might not be easily recognisable but the effects on a population might be manifested. Depending on the circumstances and need, assessments of radiation effects may have to be made at the level of the individual, population, community, or ecosystem. Such assessments may be difficult to achieve and will depend upon many factors, such as the number of individuals within a population that are affected, the nature of the different types of populations within a community, and so on. In the natural environment, the situation can become very complex because of the interaction between each individual and its surrounding ecosystem. The effects can also be modified by the presence of other environmental stressors or by combined effects related to the presence of other pollutants, and by interactions between different trophic levels.

(76) An important factor in ecology is the interdependence of populations and communities. A change in one ecological factor may have a drastic effect on another. Ecosystems consist of a certain number of biotic and abiotic components, and the radiation response depends in part on the radiosensitivity of individual biotic components prevailing in the ecosystem.

(77) Effects upon ecosystems are usually observed at the population or higher levels of organisation, whereas information on dose responses is usually obtained at the individual (organism) level. Thus there is a need for conceptual linkages between molecular effects at the individual level to potential population-level and ecosystem-level effects. These can be many and complex (Fig. 3.1), and assessment of impacts of environmental stressors, including radiation, beyond the individual level is limited by lack of scientific knowledge. It therefore seems appropriate to focus on the individual for the purpose of developing a framework of radiological assessment that can be generally applied to environmental issues, because radiation effects at the population level—or higher—are mediated via effects on individuals of that population. This approach is consistent with existing assessment methods for non-radiological environmental contaminants. Although theoretical models representing energy flow in ecosystems, predator-prey interactions, and population dynamics have been developed for a limited number of simplified ecosystems or economically

important species, there is in general a lack of data with which to assess the effects of environmental contaminants, including ionising radiation, on these important ecological functions. Consequently, assessments of the ecological effects of contaminants have usually focused on assessment of effects on individuals of the most

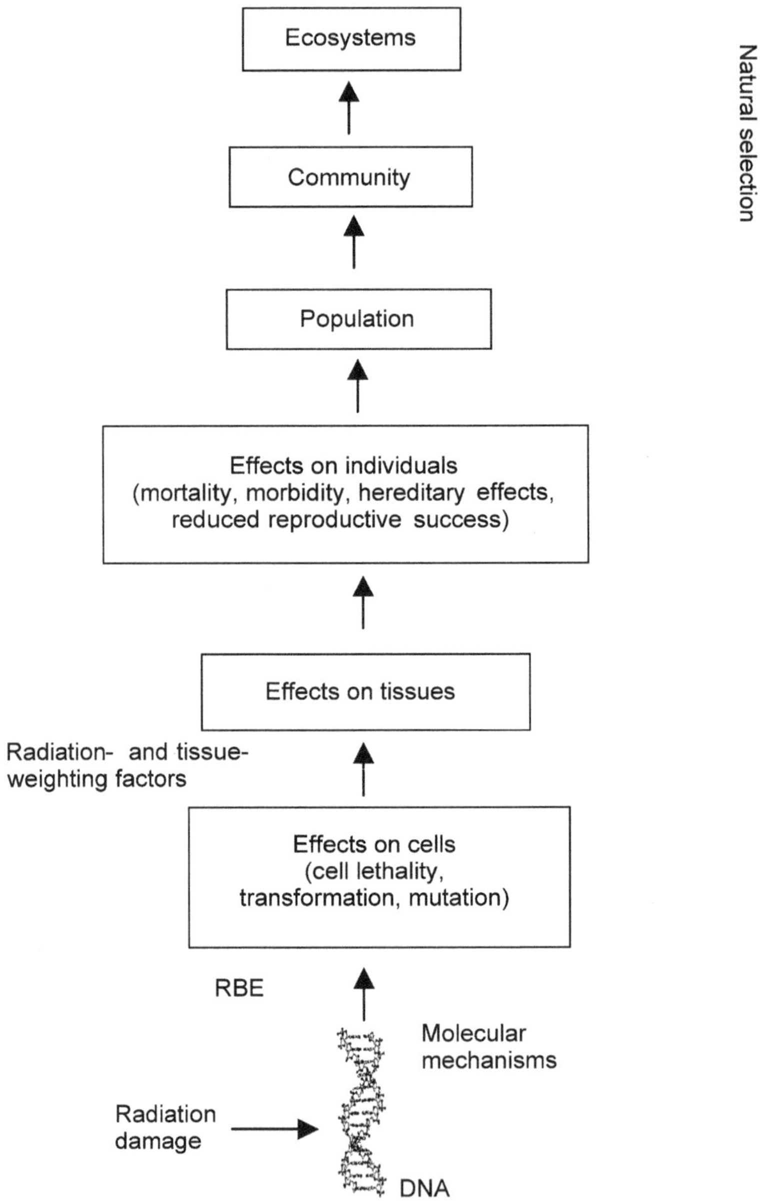

Fig. 3.1. Schematic flow chart showing radiation effects from the initial DNA damage to effects on individuals and higher levels of organisation. RBE, relative biological effectiveness.

exposed and/or most sensitive species or life stage, with the conclusion that if the most sensitive species or life stage is protected, ecosystem integrity will also be protected. Research is being conducted to support a more comprehensive approach to the assessment of radiological effects on ecosystem function (Brechnignac, 2001, 2002a,b).

(78) Although most of our information on the effects of radiation is based on studies of individuals, some field observations on populations, ecosystems, and communities have been made under controlled laboratory and experimental field conditions, and some observations are available from studies made following the accidental releases of high levels of radionuclides into the environment (IAEA, 1992; UNSCEAR, 1996; Van der Stricht and Kirchmann, 2001). Such studies have shown that reproduction is likely to be the most limiting endpoint in terms of survival at the population level, depending on the definition of what a population is and what constitutes its survival. Interpretation at a community level is, however, more complicated. Sensitivity to chronic radiation has been shown (IAEA, 1992) to vary markedly among different taxa; certain mammals, birds, reptiles, and a few tree species appear to be the most sensitive terrestrial organisms. The IAEA has also reviewed the existing literature in relation to the ICRP (1991) statements – and on the presumption that non-human species are viewed and valued more as populations than as individuals (IAEA, 1992). Various conclusions about the dose rates that would not cause observable changes to populations of generalised terrestrial plants and animals, and aquatic animals, were made and these have subsequently been used by the USDOE in a regulatory context (USDOE, 1993, 1996). Similar conclusions were drawn by a more recent UNSCEAR review (UNSCEAR, 1996). Using an ecotoxicological approach, Canadian Nuclear Safety Commission (CNSC) staff derived slightly different expected-no-effect values for use in ecological risk assessments conducted in support of regulatory requirements (Bird et al., 2003).

(79) In the long run, in several exposed communities of plants and animals, the resulting effect on the ecosystem of ionising radiation is likely to be determined by a balance between damaging and recovery processes. Effects of ionising radiation on flora and fauna are always modified by the action of a range of ecological factors. Compensatory, additive, or synergistic effects of radiation and other environmental factors may therefore be expected (Stilling, 1999).

(80) In summary, although there would appear to be much that is common about the effects of radiation on living organisms generally, there are limitations of our current knowledge about effects of radiation on non-human organisms. The immediate problem is not so much one of a lack of data, but a lack of direction in how best to organise and interpret it for the purposes of assessing impacts on non-human species. Such a re-appraisal would also greatly facilitate the derivation and prioritisation in the acquisition of new data.

4. THE COMMISSION'S SYSTEM OF PROTECTION

(81) Several basic assumptions are implied by the Commission's current statement set out in paragraph 16 of *Publication 60: 'The Commission believes that the standards of environmental control needed to protect man to the degree currently thought desirable will ensure that other species are not put at risk'* (ICRP, 1991). It is implicitly assumed that the Commission has considered the environment in terms of protecting its biological elements through its system for protection of humans, and that the protection endpoint for non-human organisms is only reproductive capacity at the population level or above. So far, the Commission has not explained how it addresses the issue of whether the environment should be protected in its own right or in the interest of humans, nor has it even explicitly stated that the environment should be protected. It can also be noted from another sentence of the same paragraph *('At the present time, the Commission concerns itself with mankind's environment only with regard to the transfer of radionuclides through the environment, since it directly affects the radiological protection of man.')* that the Commission has so far not considered all other aspects of the environment other than those directly of relevance for the radiological protection of humans. The current ICRP statement may therefore be open to different interpretations. Also, the statement itself is viewed by some as unresponsive to modern environmental concerns and sensibilities of society, and that radiological protection of the environment should be consistent with policies adopted for chemical toxins and other hazards.

(82) From a historical point of view, the anthropocentric focus of radiological protection has been prioritised because of the need to protect humans in different circumstances (medical and occupational exposures, and exposures to the public). In doing so, parts of the environment (the human habitat) probably have been afforded a fairly good level of protection through the application of the ICRP system for protection. Nevertheless, there are clearly circumstances where the Commission's current view is insufficient to protect the environment, or even incorrect. Examples are environments where humans are absent (e.g. aquatic environments), situations where humans have been removed for their own safety (e.g. in the case of intervention), and circumstances where the distribution of the radionuclides in the environment is such that the exposure to humans would be minimal, but other members of the flora or fauna could be considerably exposed. Another problem is that the implicit level of protection (i.e. not endangering whole species) is inconsistent with sustainable development and many current environmental protection policies, acts, and regulations.

(83) There are few examples where radiation exposures to both human beings and other living components of the environment have been assessed simultaneously. Where they have been, as around Sellafield in the UK, dose rates to the fauna of up to two orders of magnitude greater than dose rates to the general public (critical group) have been observed (Woodhead, 1973). Theoretical comparisons have also been made via 'desk-top' studies (IAEA, 1992), but the radionuclide concentrations used, and hence the dose rates arising from them, have usually related to the tissues and organs that are eaten by people (such as muscle tissue) rather than the tissues and organs that are of relevance to radiation effects on the organisms themselves.

4.1. Human risk assessment

(84) The ICRP system for the protection of humans is achieved in practice by the principles of justification, optimisation, and dose limitation, as well as by the use of (a) reference anatomical and physiological models of the human being, (b) studies at the molecular and cellular level, and (c) experimental animal studies and epidemiological studies. The use of models has resulted in the derivation of practical, tabulated information on the anticipated 'dose-per-unit intake' of different radionuclides that can be applied to workers, patients, and the public. The use of epidemiological and experimental studies has resulted in the estimation of risks associated with the external and internal exposure to radionuclides. For deterministic effects, the data come from human experience, supported by experimental biology. For stochastic effects (principally cancer but also including hereditary effects), the ICRP's starting points are the results of epidemiological studies. These are supplemented by information from experimental studies on the mechanisms of carcinogenesis, in order to provide risk estimates at low doses of interest in radiological protection.

(85) The ICRP's risk estimates are called 'nominal' because they relate to the continuous exposure of a nominal population of females and males with a typical age distribution. As with all estimates derived from epidemiology, the nominal risk coefficients do not apply to specific individuals, unless it can be assumed that the individual is typical of the nominal population. If one accepts these assumptions, the estimates of fatality and detriment coefficients are adequate both for planning purposes and for general prediction of the consequences of exposures of a nominal population. For estimation of the likely consequences of an exposure of an individual or a known population, it will typically be better to use absorbed dose and specific data relating to the RBE of the radiations concerned, and estimates of the probability coefficients relating specifically to the exposed population or individual.

(86) The ICRP system for assessment is robust and is, in several aspects, in conformity with what is used in other fields of environmental protection, e.g. the identification of hazards (essentially all radionuclides), risk identification (primarily through DNA damage), and risk characterisation involving reference values. However, this system for assessment does not apply to the environment.

4.1.1. Reference Man

(87) Calculations of radiation dose to an organism from external or internal sources require information about the anatomical and physiological characteristics of the exposed organism. In order to have consistent and reproducible radiological protection guidance for different types of exposures, it is important that a consistent set of reference values be used to describe, prospectively, various anatomical and physiological characteristics of an exposed individual. These reference values for tissues and organs, when summed, define a reference individual. Consideration of an entire reference individual helps to ensure that there will be internal consistency about how the volume, mass, or functional characteristics of various organs or tissues are specified.

(88) This concept of a primary reference organism for human radiological protection (Reference Man) has long been used and recognised by the ICRP. The work to define the first ICRP reference individual began in the late 1940s, and in 1975, *Publication 23* (ICRP, 1975) on Reference Man was published. This report contained a wealth of information on the anatomical, morphological, and physiological characteristics of humans related to the biokinetics or dosimetry of internally deposited radionuclides. The ICRP has recently adopted a new report that provides up-dated information on Reference Man (ICRP, 2002).

(89) Reference Man is not intended to describe an 'average' individual of a specified population group, nor do the data sets of this reference individual necessarily represent data that would be obtained by taking a random sample of any particular population. The purpose of Reference Man is to create a 'standard', and a point of reference, for the procedure of dose estimation to humans. The parameters and characteristics were originally defined in order to provide a basis for estimating exposures to workers, and have with time been complemented with subsets of the primary Reference Man, such as the Human Respiratory Tract Model (ICRP, 1994), the skeleton (ICRP, 1995), and doses to the embryo and fetus from intakes of radionuclides by the mother (ICRP, 2001a).

4.2. Revision of the Commission's existing recommendations

(90) ICRP is at present revising its existing recommendations for the protection of humans in order to develop recommendations for the beginning of the 21st century (Clarke, 1999; ICRP, 2001b). In doing so, the Commission is considering a simplified approach, with the following major changes from the recommendations in *Publication 60* (ICRP, 1991):

- an emphasis on the protection of individuals, in addition to the requirement to optimise protection;
- a broadening of the narrow definitions of dose limits to a range of protective actions and the level above which each action should be taken;
- a distinction between protective actions that can be applied to the source and those that can be applied only to the pathways leading from the source to the doses in individuals;
- the inclusion of a policy for radiological protection of non-human species; and
- the clarification of the dosimetric quantities.

This is, therefore, a good opportunity to include a framework for assessment of radiological impacts on non-human species that could be interfaced with, or integrated into, the system of radiological protection.

(91) A scale indicating the appropriate level of concern has been suggested by ICRP (2001b). The aim is to specify a broad basis for defining levels of concern and to avoid a rigid demarcation of the bands while avoiding ambiguity. There are several factors that influence the choice of these levels of concern. The global average

effective dose from natural sources excluding radon is about 1 mSv/year, and the Commission has proposed that this dose could be a starting point. The natural background provides no justification for additional exposures, but it can be used as a basis of judgement about the importance of other exposures.

4.3. The need for reform

(92) Radiological protection of the environment is an important issue and will be even more so in the future. The human-orientated approach used up until now by ICRP has obvious limitations with respect to the biosphere as a whole. The current system of radiation protection is not generally applicable to the environment, nor does it correspond to managerial needs or society's demands. The Commission's current policy statement, expressed in paragraph 16 of *Publication 60* (ICRP, 1991), is increasingly being challenged because of its lack of supporting evidence, transparency, and its lack of connection with society's environmental protection objectives, as stated in the principles presented in Chapter 3. Therefore, it is necessary that the Commission formulates a more comprehensive approach to embrace the protection of both humans and other living organisms. In doing so, the two most important questions are as follows.

- Can the current ICRP system of radiological protection be extended to protect biota?
- How can the ICRP recommendations for the 21st century be designed so that they will also explicitly include consideration of possible impacts of radiation on species other than humans?

(93) In answering these questions, it is important to recognise that it is not the role of experts in radiological protection to define what parts or segments of the environment need protection from radiation. The needs and objectives for protection of the environment have already been defined by society at regional and national levels. The role for ICRP will be to interpret the consequences for radiation protection of the shift towards a more comprehensive approach, and to define how the Commission's long experience and systematic approach in the radiological protection of humans can contribute to the achievement of these objectives.

(94) Bearing in mind the fact that the effects of ionising radiation, at least at the molecular level, are similar for all living organisms, there are many reasons why ICRP now needs to consider the basis upon which it can state more clearly its position and future role with regard to the protection of non-human species. These reasons include:

- the need to demonstrate that the principles of radiological protection are consistent with a recognition that it is essential to consider the interdependence of humans and the environment in order to achieve sustainable development;
- thc necessity for operators and regulators to demonstrate compliance with the increasing number of existing international and national environmental

requirements pertaining to those practices that release radionuclides into the environment;
- the historic and continuing use made of ICRP recommendations and advice in the derivation and formulation of both international and national law pertaining to the regulation of such practices for protection of human health;
- the need of national bodies to provide advice with respect to intervention situations, particularly where the potential for human exposure is either minimal or preventive action has already been undertaken;
- recognition of the necessity to demonstrate explicitly how knowledge of the potential effects of ionising radiation on the environment can be used to inform decision making and the public; and
- the need to bring the basis for the regulation of exposure to ionising radiation, in an environmental context, more in line with the regulation of other potentially damaging industrial practices or of other contaminants associated with practices of interest to ICRP.

(95) The role of the Commission in this matter would of course be greatly facilitated if there existed a single ethic that encapsulated what was meant by *protection of the environment*. But there is not. Concern for the environment arises from many quarters, and the basis for its protection, as discussed in the preceding sections, can often be traced back to:

- scientific evidence, particularly with respect to the need to protect those aspects of the environment that directly or indirectly affect human health, human sustenance, or human wealth and livelihoods;
- social and cultural concerns, which may in turn have their basis in religious or philosophical tenets and beliefs; and
- the need to comply with international and national law that has arisen with respect to the protection and conservation of the natural environment.

(96) Taken individually, any of the above reasons for the Commission's involvement in this subject could be debated at length, but collectively they make it difficult for ICRP *not* to get involved. But equally, ICRP should not derive an ethic upon which protection of non-human species should be based in isolation from what has already been done both nationally and internationally. There is, however, sufficient evidence to indicate the level of interface required between a knowledge of radiation effects on the one hand, and the requirements of protection of biota on the other—namely, the needs of environmental conservation, the maintenance of biodiversity, the meeting of environmental quality objectives, and the requirements of ecosystem health.

(97) Given the speed with which this subject is developing nationally and internationally, and the lack of international adoption of any existing systematic and structured approach to assess and manage radiation effects on non-human species, there are strong expectations from many quarters for ICRP to act. It is therefore clear that the Commission must define its position and desired future role as regards radiological protection of biota in order to meet the expectations from other international bodies.

5. PROPOSAL FOR A SYSTEMATIC APPROACH TO ASSESSING RADIOLOGICAL IMPACTS ON NON-HUMAN SPECIES

5.1. Introduction

(98) The Task Group was specifically asked to identify or suggest a basis for a framework that could be used by the Commission to help inform the further development of environmental protection approaches at regional or national level. In doing so, it was necessary to first review briefly the current state or recent advances in approaches for assessing and managing environmental risks attributed to exposure of ionising radiation arising from human activities. Some of these reviews have been developed to meet national needs, others have been more general in concept and origin; some have been specifically developed in order to address the perceived problems inherent in the ICRP statements, whilst others have sought to approach the problem afresh, from first principles.

(99) Thus a number of different approaches have been made to address the questions raised with respect to the current ICRP statement on environmental protection. They include the following:

- arguments that because humans are an integral part of 'the environment' and are afforded such a high level of protection, all other components of it are axiomatically protected;
- calculations to demonstrate that, in hypothetical situations, if radionuclide concentrations in the environment are such that the 1 mSv/year dose limit to humans is not exceeded, the concentrations of radionuclides in the animals and plants in their food chain would therefore receive dose rates less than those likely to cause them 'harm' at the population level (IAEA, 1992);
- the use—or proposed use—of **dose-limit standards**, for the protection of populations of all aquatic animals (1 rad/day) and consideration of dose 'standards' of 1 and 0.1 rad/day for populations of all terrestrial plants and animals, respectively, for certain sites managed by the USDOE (USDOE, 1993, 1996; UNSCEAR, 1996), and for marine animals (100 mGy/year) and marine plants (1000 mGy/year) in Russia (Sazykina and Kryshev, 1999);
- the introduction of an **ecological risk assessment framework** to assess the effects on non-human species of radionuclides released from nuclear facilities using dosimetric models and estimated 'no-effect dose rates' for a number of biotic assessment endpoints relevant to aquatic and terrestrial ecosystems (Bird et al., 2003; Thompson et al., 2003);
- an attempt to develop an overall system for environmental protection based on a narrowly defined **reference fauna and flora** approach consisting of defined dose models, data sets to estimate exposures, and data on dose–effect relationships for individual fauna and flora that could be used to help decision making (along with other relevant biological information) in different circumstances, such as control of practices or in cases of intervention (Pentreath, 1999, 2002, 2003), which has been supported by IUR (2000); and

- consequent developments to produce **systematic frameworks**—also using a reference fauna and flora approach—for assessing environmental impact of ionising radiation in specific geographic areas, such as at national levels (Coppleston et al., 2001), and for European and Arctic ecosystems, including projects financed through the EC 5th Framework Programme, notably FASSET and EPIC (Strand et al., 2000).

(100) All of these approaches have their strengths and weaknesses. Criticisms raised with regard to the first (the 'axiomatic') approach include the fact that, even if humans are present, they are unlikely to receive the highest doses because of the spatial distribution of radionuclides in the environment, and because of the differences in the biological accumulation of radionuclides by different fauna and flora; plus the fact that there are sectors of the environment where humans cannot live (underwater) and circumstances where they have been removed for their own safety (intervention) but the fauna and flora remain. The last two of these criticisms can also be levelled at the 1992 IAEA study, plus the fact that the organisms assessed form part of the human food chain, and may not represent the organisms that may be the most exposed to radionuclides. But perhaps one of the greatest weaknesses inherent in both of these approaches is that the level of protection sought for, or afforded to, the environment is not sufficiently defined in terms of biological endpoints or the levels of risk associated with them; although the IAEA study was centred on effects at the 'population' level without defining what it meant by population.

(101) The implementation of **'dose-limit standard'** approaches requires consideration of the following: by what 'agreed' methodologies were the limits derived; what biological endpoints—or levels of risk relating to them—do they represent; how does one demonstrate compliance with them and how often; and what does one do if they are exceeded? With regard to the USDOE's existing and proposed dose limits, these 'expected safe levels' of exposure are based on published data on acute and chronic radiation effects (NCRP, 1991; IAEA, 1992; UNSCEAR, 1996), with reproduction being the critical endpoint of concern, and based on the assumption that the population will be protected adequately if the dose rate to the maximally exposed individual does not exceed that level of exposure. The USDOE's dose limits are not applied in a way such that their exceedance would require a mandatory regulatory or remedial action, but rather as dose-rate guidelines that, if exceeded, provide an indication that further investigation and action is likely to be necessary. A graded approach for evaluating radiation dose has been developed as a means of demonstrating compliance with dose limits, and as a tool for conducting screening assessments of radiation impact. A USDOE Technical Standard documents the relevant methodology, provides guidance on frequency evaluations, and provides guidance on how to proceed if the dose limits are exceeded (USDOE, 2002). The application of four generalised organism types (aquatic animals, riparian animals, terrestrial animals, and terrestrial plants) in the derivation of limiting concentrations of radionuclides (Biota Concentration Guides) in soil, sediment, and water as a general screening tool is described by Higley et al. (2003a,b,c).

(102) The **ecological risk assessment framework** also uses a tiered approach to determine whether or not actual or planned releases of radionuclides may be harmful to biota (CEPA, 1999 and CEAA, 1992 provide definitions of 'harm'). Measurement endpoints, based on appropriate reference species, are chosen on the basis of a pathways' analysis, and available scientific data on the relative sensitivity of various taxonomic groups and on ecosystem functions to ionising radiation. When the results of such assessments indicate a potential for harm, they are linked to a risk management framework in which available mitigation measures are considered in a cost-benefit analysis. The success of the chosen mitigation measure is then evaluated against environmental performance objectives using environmental monitoring and/or modelling techniques.

(103) Finally, the difficulty with attempting to develop the **reference fauna and flora** system approach of Pentreath (1999, 2002, 2003) is seen to be the potential scale of the task, the extent to which a 'reference' approach based on a few well-defined reference organisms could usefully be applied to many different specific locations or circumstances, and the individual basis — not taking account of impacts on higher levels of organisation. The IUR has supported the reference fauna and flora approach and integrated the concept into their ongoing development of an environmental protection framework (Strand et al., 2000; IUR, 2000). A rather basic reference fauna and flora approach to establish release rate limits was, however, first used with respect to evaluating potential environmental impacts of radionuclide releases into the marine environment (Pentreath and Woodhead, 1988), and this was applied by the IAEA in its consideration of re-defining annual release rate limits for the purposes of the London Convention (IAEA, 1988). Both the USDOE and the Canadian approaches also make use of some form of generic reference 'organisms' or entities for assessing compliance with predefined dose-rate limits or for calculating doses to exposed organisms to calculate a risk quotient by comparison with dose–effect data on relevant endpoints. The more recent developments of 'reference' approaches are therefore considered in this document in some detail because their practical application is also an integral part of current assessment and management framework development approaches in many countries — the EC's FASSET programme involving Finland, France, Germany, Norway, Spain, Sweden, and the UK; the EPIC programme involving Norway, Russia, and the UK; and the UK's Impact Assessment programme (Copplestone et al., 2001).

5.2. Objectives for protecting non-human species

(104) Much discussion has taken place with regard to defining what one is actually aiming to assess—risks to individuals, populations, or ecosystems—by the application of any of these approaches. In this respect, the situation is often compared—or contrasted—with radiological protection of humans, where the aims are (relatively) clear. To some extent, however, the need to answer such a difficult question is becoming increasingly less because of the emergence of requirements that have to be met with respect to a growing body of general and specific environmental legislation. Thus at international and national level, a growing list of animals, plants, areas,

habitats, and so on are afforded protection in law from 'harm' at the individual or population level, as variously described, from all manner of activities including the release of radionuclides. Furthermore, many international conventions and national legislations focus on pollution prevention regardless of the level of risk to non-human species.

(105) Equally, however, as discussed in Chapter 2, it has to be accepted that there is no single ethic that encapsulates what is meant by 'environmental protection'. There is, therefore, no real context for asking the question: what are we, the radiation protection community, trying to protect? Such protection as is currently afforded internationally has arisen via a miscellany of global and regional agreements that relate to pollution control, waste management practices, hazard minimisation, and the need to conserve and protect the natural environment and individual components of it. Nevertheless, collectively, this complex web of multilateral environmental and similar agreements already constrains a large number of industrial practices worldwide. These international agreements are all effectively 'soft' laws in that they are not, generally, strictly enforceable — unless they apply across a number of member states within a broader political and legal framework. Implementation is therefore usually via national legislation. At a national level, more specific legislation is applied in relation to all of these subject areas, particularly with regard to the use of technologies in delivering the objectives of pollution control. This hierarchical approach is also relevant to the level and extent to which international advice can sensibly be given to protect the living environment from harmful effects of ionising radiation.

(106) Many international and national agreements now require that environmental risk assessment be accomplished in a transparent fashion, that is, in an iterative and reproducible fashion, through environmental impact assessments. The requirement is how best to demonstrate compliance with all of the existing and forthcoming 'environmental protection' legislation relevant to that site or practice. This may take the form of essentially having to prove a negative—that practices do not cause or result in harm to the environment—or that emissions from practices are harmless. With regard to protecting the living components of the natural environment in terms of 'nature conservation', the requirements are again—as noted in Chapter 2—usually to *conserve* particular species or habitats; to *maintain the diversity* of habitats, of species, and of the genetic variability within species; and to *protect* habitats and designated areas that are from time to time identified for one reason or another.

(107) One relevant general question is therefore: what effects of radiation on fauna and flora would have to be minimised in order to meet such requirements? The answer is, quite clearly, very many. In order to simplify matters, and to enable the development of a useable management framework, Pentreath (1998, 1999, 2002, 2003) and the IAEA (IAEA, 2002b) have suggested that a suitable interface would be that of summarising such effects into three broad categories: early mortality directly attributable to radiation; scorable cytogenetic (DNA) damage—as an indicator of undefined biological damage; and reduced reproductive success. A fourth endpoint could be morbidity related to radiation damage. In taking this approach, however, it

was fully recognised that such headings mask a large variety of discrete effects—such as those relating to fertility, fecundity, and so on. But, equally, it also fully recognised the limitations of our current knowledge of such effects, plus the need to be able to interpret such expected effects within a broader context of managerial evaluation of impacts on the environment. Thus, for example, there may be different consequences—and thus decisions to be made—if only a small fraction of a natural population were to be exposed to a 'high' level of dose, compared with a large fraction of the population being exposed to a 'low' level of dose. However, defining what constitutes an acceptable level of harm goes beyond the realm of science and is best dealt with at the environmental management stage when policy decisions take socio-economic factors into account.

(108) If any assessment system is to be of value, it also needs to be capable of application to any managerial circumstance. In the current terminology of radiological protection, this effectively means that it could be applied to both *practices* and *interventions*. More specifically perhaps, it could be used where concern is often uppermost in the minds of policy makers and the general public: in situations where pathways to humans are few or entirely absent; in predicting future exposures, especially with regard to waste disposal practices; in situations where accidents may happen or things do not go as predicted; in dealing with contaminated land which is unlikely to be inhabited by humans; and in providing explicit assurance for routine operations. Fortunately, a considerable effort has been expended in monitoring actual releases, and in modelling future or possible releases, of radionuclides in a wide range of terrestrial and aquatic environments. Many studies have been made of the relative concentrations of radionuclides likely to occur between the different media and the fauna and flora that live in them, particularly for aquatic environments. Unfortunately, however, most of these data—usually expressed as concentration factors, concentration ratios, or transfer factors—are for either whole organisms or those parts of organisms that are likely to be eaten by humans. In other words, they are not always the data that are required to estimate the dose rates to fauna and flora that would have a bearing on early mortality, reduced reproductive success, or hereditary effects.

(109) Provided that one is willing to rely largely on models instead of on direct empirical measurements, a large number of models and relevant databases do exist, and attempts have been made to develop 'reference' ecosystems for particular radioactive waste disposal practices — such as the IAEA's BIOMASS (Biosphere Modelling and Assessment) programme (Linsley and Torres, 2001). The reference fauna and flora approach would therefore be readily applicable to a variety of environmental situations. For the marine environment, concentration factor and k_d values that could be used as a basis for calculating dose-per-unit-water concentration (via both internal and external exposure) for a number of faunal types have already been compiled for some 60 elements (IAEA, 1985). Such an approach was used by the IAEA for modelling exercises with regard to the practice of sea disposal for some 200 radionuclides (IAEA, 1988). Such data sets would then be, effectively, the equivalent of tabulations of dose-per-unit-intake for Reference Man.

5.3. A reference fauna and flora approach

(110) As can be seen, there have been several 'reference' approaches, where the term 'reference' refers to different things — dose models, methods, and so on. However, the reasoning behind the development of a systematic reference fauna and flora approach (Pentreath, 1998, 1999, 2002, 2003; Pentreath and Woodhead, 2001) was to derive a reasonably complete set of related information for a few types of organism that were typical of the major environments. This would be achieved by drawing upon our existing information on them, and on the basis that such organisms would be amenable to further study in order to obtain a more complete understanding of a few basic aspects of their responses to radiation that were important in an environmental protection context. It therefore essentially admits that this approach cannot provide a general assessment of the effects of radiation—or indeed of any other environmental contaminant—on the environment as a whole. But, by using reference data sets, one should be able to make some sort of statement about the probability and severity of the likely and different effects of radiation exposure on such individuals. One should then, in turn, be able to make an assessment of the likely consequences for either individuals or the relevant population, using these and other environmental data and information, in order to make managerial decisions relevant to the circumstances that gave rise to the radiation exposure.

(111) This concept of deriving such data sets for reference fauna and flora is therefore similar to that of the reference individual (Reference Man) used for human radiological protection, in that it is intended to act as a basis for many calculations and decisions. It is also similar to the concept of assessment and measurement end-points used in the Environmental Risk Assessment (ERA) frameworks (Suter, 1999). It is intended that each reference organism would, as for Reference Man, serve as a *primary* point of reference for assessing risks to organisms with similar life cycles and exposure characteristics. More locally relevant information could be compiled for any other fauna and flora, but each such data set would then have to be shown to be related in some way to the reference organisms. The choice of primary reference organisms for flora and fauna will depend on the future development of environmental protection from radiation. Parameters for choosing reference organisms could include the role of the organism, its radiosensitivity, and the role of the organism in the ecosystem.

(112) Thus, for each reference fauna and flora for environmental assessments (Pentreath, 2002, 2003), one should have (or be reasonably able to obtain) a fairly internally consistent set of data on the following: basic life-cycle biology; pathways of exposure to radiation that can be expressed in terms of dose-per-unit-exposure ('look-up' tables); dose model(s) to estimate doses received by the relevant 'critical' organs; and effects of radiation (early mortality, reduced reproductive success, and observable DNA damage) on individuals. Such data sets, for a number of reference organisms, would also serve as 'default' values for use in various assessment scenarios.

(113) This still raises the question of what these *primary* reference fauna and flora actually are (Pentreath and Woodhead, 2001). The criteria upon which they might

be selected would probably include many scientific considerations, but it is equally important to have regard for the extent to which they are considered to be *typical* representative fauna or flora of particular ecosystems and particular exposure pathways. Ideally, one might like to select those organisms that were known to be particularly sensitive to radiation, or were known to be vital components of particular ecological communities or expected to receive higher exposures because of their habitat (e.g. sediment-dwelling organisms when radionuclides will accumulate in sediment). But one also has to be pragmatic, and therefore consider the amount of radiobiological information that is already available on them, including data on radiation effects. They would also have to be amenable to future research in order to obtain the necessary missing data. One would also have to consider the extent to which they have some form of public or political resonance, so that both decision makers and the general public at large are likely to know what these organisms actually are, in common language — such as a duck or a crab. By and large, one could argue that a starting point could be those types of fauna and flora for which we already have data on doses and radiation effects. But it does not matter if any properties (dose received, radiation effect) of other fauna and flora are already known to be greater or less than that of the 'reference' ones, providing that the scale of the relationship to them is known. Once selected, the fauna and flora would still need to be described in taxonomic terms. It has been suggested that 'species' level is probably too narrow, and that 'family' or 'order' level might be an appropriate level to start aggregating existing data (Pentreath and Woodhead, 2001).

5.3.1. Dose models and exposure geometries

(114) The variety of dose models needed for such reference organisms, in addition to the obvious considerations of target size and shape, will clearly depend upon how the consequences of radiation result in one of the above categories of biological effect. A short hierarchy of dose-model complexity has been suggested by Pentreath and Woodhead (2001) based on a solid sphere, ellipsoid, or cylinder, plus one (or more) internal solid sphere(s), ellipsoid(s), or cylinder(s) to represent other specific tissues of interest. Such models have been used extensively in the past (Woodhead, 1979; IAEA, 1988; Pentreath and Woodhead, 1988; NCRP, 1991); each has its advantages and disadvantages. Many of these models are providing the basis of the current studies being made within FASSET (Copplestone et al., 2001) including comparisons using Monte Carlo simulations. Variations of these environmental geometry categories are also already commonly applied (Amiro, 1997; Jones, 2000; Higley et al., 2003a,b,c; USDOE, 2002). An equally important consideration is that of the possible range of 'environmental' geometries within which these dose models could be set. For convenience, it has been further suggested that these could be grouped into the following simple categories: surrounded by air, water, or soil (4π); at the interface of air or water with soil or sediment (2π); and concentric, i.e. organism surrounded (4π) by air or water and then surrounded by soil or sediment (Pentreath and Woodhead, 2001).

5.3.2. Dose consideration levels for non-human species

(115) A further obvious question that arises with regard to the reference fauna and flora approach is that of how to interpret and apply data on the various relationships between different doses and different biological effects. There are several ways in which environmental assessments can be made by using measured or estimated dose rates. Comparisons can be drawn with the range of natural or historic background radiation levels obtained in a given area, including internal exposure for specific species. They can also be compared with experimentally derived information on what harmful effects have occurred at what levels of radiation (doses and dose rates). Both of these approaches have been used (Amiro and Zach, 1993; UNSCEAR, 1996; Bird et al., 2003). For the protection of the public, the ICRP is now considering an approach based on *levels of concern*, and with explicit reference to background dose rates (ICRP, 2001b).

(116) This idea also lay behind the proposal to have *derived consideration levels* for fauna and flora — where data derived could be set out in scales of dose-effect levels in order to aid in the *consideration* of different management options, depending upon exposure circumstances and all other relevant information (Pentreath, 1999, 2002, 2003). But, in this case, there are currently only two bases upon which to assess the potential consequences for fauna and flora: natural background dose rates and dose rates known to have specific biological effects on individuals. This approach would, of course, place considerable reliance on knowing what the mean and range of background dose rates actually are; hence the reason to be clear as to how to describe them in terms that allow for the high alpha content found in many aquatic organisms.

(117) Bands of *derived consideration levels* for reference fauna and flora could be compiled by combining information on logarithmic bands of dose rates relative to normal natural background dose rates, simply as a means of presentation, plus information on dose rates that may have an adverse effect on reproductive success, or result in early mortality (or cause morbidity), or are likely to result in scorable DNA damage for such organisms. Such a banding could be essentially on the same basis as that recently proposed for humans (ICRP, 2001b), in that additions of dose rate that were only fractions of *their* background might be considered to be trivial or of low concern; those within the normal background range might need to be considered carefully; and those that were one, two, three or more orders of magnitude greater than background would be of increasingly serious concern because of their known adverse effects on individual fauna and flora (Pentreath, 2002). This approach has the added advantage of not relying on the results of arbitrary extrapolations of effects on individuals to effects at higher levels of organisation (e.g. population, community). The Task Group refers to the background radiation relevant to a given non-human organism. Natural background is often dose rate measured in air (UNSCEAR, 1996), whereas the Task Group refers to the absorbed dose in that organism.

(118) Other factors would also have to be taken into account, particularly with regard to ethical, legal, and social considerations, and the nature and numbers (or

fraction of the local population) of fauna and flora that were liable to be exposed within the different bands; in other words, all ethical, legal, and social factors as well as the scale of the area actually or likely to be affected in terms of elevated dose rates, and the specific nature of the fauna and flora that lived within it. This would effectively define the boundary between radiation protection expertise, other areas of biological and ecological sciences, and social issues such as stakeholder rights and democratic procedures. For a reference faunal type such as a terrestrial mammal, the result might be similar to that outlined in Table 5.1 (adapted from Pentreath, 2002).

Table 5.1. An example of how a table of derived consideration levels might look for the case of a reference terrestrial mammal (as modified from Pentreath, 2002)

Derived consideration level	Relative dose rate (incremental annual dose)	Level of concern
Level 1	< Background	Low concern. No action considered.
Level 2	Background range	Low concern. No action considered.
Level 3 and higher	> 10 times background	Concern dependent upon the nature of effects, the numbers and types of individuals affected, the spatial and temporal aspects, and so on. Remediation may be considered at extremely high relative doses.

5.4. Developing a common approach to protect humans and other living organisms

(119) It is necessary that a system for radiological protection of non-human organisms is harmonised with the principles for the radiological protection of humans. The objectives of a common approach to the radiological protection of humans and other living organisms, as suggested elsewhere (Pentreath, 2002), might be **to safeguard human health** by preventing the occurrence of deterministic effects and limiting stochastic effects in individuals and minimising them in populations; and **to safeguard the environment** by preventing or reducing the frequency of effects likely to cause early mortality or reduced reproductive success in individual fauna and flora to a level where they would have a negligible impact on conservation of species, maintenance of biodiversity, or the health and status of natural habitats or communities.

(120) The development of such a common approach to radiation protection in general would also further justify the development of a common methodology and scientific basis for making assessments and decisions. Thus the achievement of these objectives should be centred on a set of reference dose models, reference dose-per-unit-intake and external exposure values, plus reference data sets of doses and effects for both humans and fauna and flora. This would support informed policy and management decision making with regard to public health and environmental risks for the same environmental situation (Fig. 5.1).

(121) Any system of radiological protection for humans and other living organisms should, however, recognise that society has established environmental protection

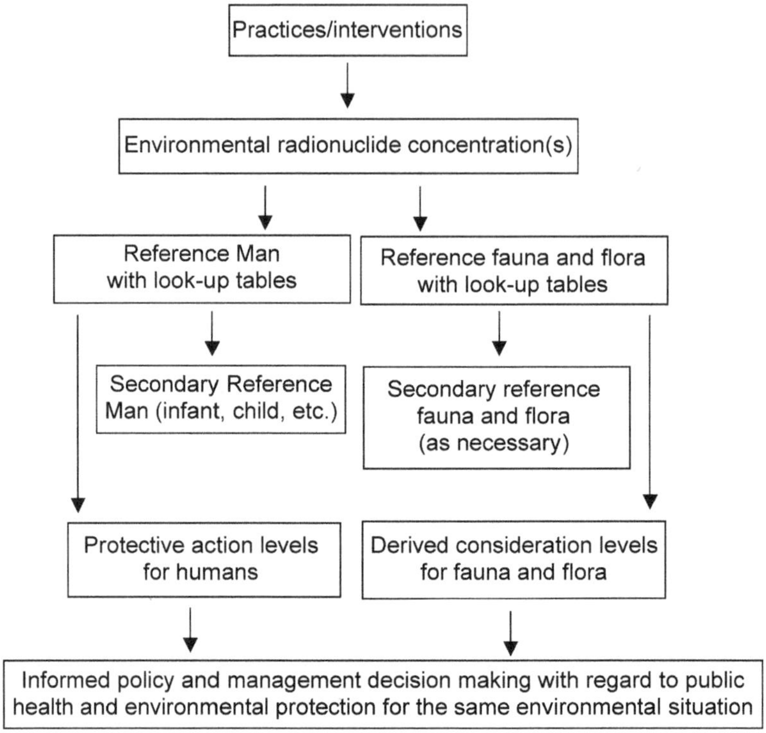

Fig. 5.1. Developing a common approach for the radiological protection of humans and non-human organisms (slightly modified from Pentreath, 2001).

goals, such as pollution prevention and waste minimisation, that may go beyond the minimisation of dose. In any case, the development of a scientifically rigorous radiological risk assessment framework for non-human biota is needed to support the many existing international conventions, national legislations, and cost–benefit analyses. In addition, media concentrations may be used as direct indicators of environmental contamination and as management tools to track the success of pollution prevention initiatives.

6. DISCUSSION

6.1. Assessment and management aspects

(122) It has to be acknowledged that within any overall approach to environmental protection, the science-based assessment of environmental consequences will be affected by—and indeed guided by—ethical and democratic decisions made by society in general. These decisions will therefore reflect the particular cultural environment of that society, expressed in terms of their moral values, as well as by managerial principles. There is, therefore, not always a clear distinction between what one might call 'purely scientific' and 'purely value-based' judgments, because science and societal views are interlinked, as discussed in Chapter 2.

(123) For the purpose of developing a system for assessing the harmful effects of ionising radiation on non-human species that will support management decisions, it is therefore useful first to differentiate between these *assessment* and *management* components. This is particularly important when attempting to understand the purpose of the analysis, because each component may use completely different methodologies and interpretations. Thus, for example, the assessment component may require an indication of general environmental consequences, whereas the management component may require demonstration of compliance with specific procedures, dose limits, or media concentrations. The difference between these two components is also reflected in the current state of the development of environmental protection frameworks within different countries, or being carried out by different organisations.

(124) With respect to ionising radiation, *assessments* may be performed in order to analyse the consequences that the presence of radionuclides may have for the environment generally, or specifically in relation to particular circumstances or ecosystems. This may require an *'effects-analysis'* approach that needs to be targeted at the correct hierarchical level of biological organisation, and covers a sufficiently wide range of biological effects and rationales for assessing their consequences at these different levels. The IUR has supported this type of assessment and made it a key component of its current work relating to the development of an environmental protection framework (IUR, 2000; Strand et al., 2000). Such an approach is being taken by the EC's 5th Framework Programme projects FASSET and EPIC, where 'effects databases' are being assembled to be used, respectively, for major European ecosystems generally and for the Arctic environment. The results of such assessments are essentially 'open ended' because they are designed to facilitate communication on environmental risks and to help the overall decision-making process, although they may subsequently also help in deriving more specific environmental criteria or standards. A number of possible *management* options may also then have to be considered in the light of the results of such assessments.

(125) A *management* requirement can, for example, involve demonstration of compliance with specific standards that are based on *assessments* and risk evaluation. Estimation of standards will also involve balancing the potential risks against possible benefits of specific actions. The procedures necessary to do this can be

simple or complex, and tiered approaches can be introduced to facilitate their use. The graded-approach system developed by the USDOE (using dose-limit standards) and which is used at some USDOE sites and facilities (Jones, 2000; Higley et al., 2003a,b,c; USDOE, 2002), and the approach recommended to the CNSC by its Advisory Committee on Radiological Protection (ACRP, 2002) based on aspects of the tiered approach of Environment Canada's ecotoxicological risk assessment, plus the graded approach of the USDOE for some of its contaminated sites, are all illustrative of systems developed, or under development, as such compliance tools. These approaches and methodologies also provide data and information relevant to other managerial requirements and assessments.

(126) Differences between the various methodologies are usually more apparent than real, and mainly depend on the *purpose* of the assessment rather than fundamental differences in philosophy or approach.

6.2. Developing a common approach to the protection of humans and non-human organisms

(127) The challenge for ICRP, therefore, is to re-examine its existing system (and its proposals) to see if it can incorporate the ideas being developed with regard to environmental protection in such a way that it provides a means of underpinning these existing and developing assessment and management initiatives. The Commission is well placed to accept this challenge because although there are clear differences between the ethical, conceptual, and practical aspects of protecting humans and non-human organisms with respect to ionising radiation, there are also many similarities. Much of the basic information on the mechanisms by which radiation can affect living matter has been derived from studies on organisms other than humans. Equally, human-derived data can help in the development of a systematic approach to protecting biota. Indeed, there are clear advantages in developing the science base of information on all aspects of the effects of radiation such that it can be applied to living matter generally.

(128) There are also good *practical* reasons for ensuring that the development of any approach to the protection of non-human species is consistent with the evolving system of the protection of humans in an environmental context. This will avoid the development of one system undermining the other, and enable both to be carried out within the same overall assessment and management framework. An over-arching, systematic approach is therefore needed in order to provide this high-level advice and guidance. This should include the following elements for both humans and biota: a clear set of principles and objectives; an agreed terminology — particularly with regard to quantities and units for biota; key reference dose models and related data sets to quantify exposures; authoritative analyses of categories of radiation effects data relevant to the needs of environmental and human health assessments; guidance on the practical application of the system; and clear ownership and management of review and revision processes in the light of new data and interpretations.

(129) It is therefore proposed that, as a starting point, ICRP takes the responsibility for defining and developing sets of assessment data for a small number of

reference fauna and flora. These would essentially be analogous to Reference Man for human radiological protection, and their number may increase as more knowledge becomes available. The purpose and objective of the primary set would be to obtain as complete a database and understanding as possible of the basic biology and the doses that could be received by a limited set of faunal and floral types, and the resultant effects of radiation on them. The criteria for the choice of reference organisms will have to be decided by the Commission with guidance from existing ecological risk assessment frameworks. The data sets, and advice on how to use them, would be for a few organisms that are typical of different natural environments, representative of various exposure pathways and of key ecological functions.

(130) It is, however, unlikely that the sole use of such a limited set of reference organisms would serve to satisfy all assessment needs (Pentreath, 2003). Therefore, the ICRP's reference set could be supplemented or supported by information on other reference organisms where, for example, there is a need for a greater overall range of faunal and floral types of organism in the assessment exercise. Such additional sets may include locally characteristic types of fauna and flora for particular ecosystems, defined either (a) in terms of habitats (forests, freshwater lakes) or (b) with respect to particular geographic regions or areas (e.g. the Arctic, or temperate Europe); or very specific faunal or floral types (e.g. in order to satisfy or comply with specific 'nature conservation' legislation). The derivation of all other assessment data sets would benefit greatly by being able to demonstrably relate to the ICRP set.

(131) One possible advantage of this approach is that for any given spatial and temporal distribution of radionuclides, from any source and under any circumstance, one should be able to estimate both the relevant *levels of concern,* with respect to members of the public (based on Reference Man), and the *derived consideration levels*, with respect to non-human species (based on reference fauna and flora). These two '*bands*' would be independent of each other but derived in a complementary manner, based on the same underlying understanding of the effects of radiation on living matter. Also, in a practical sense, the two '*bands*' would (or could) each be related to the same concentration of a specific radionuclide, within a specific environmental material, at any particular site (Fig. 5.1). Thus, for example, a given concentration of one radionuclide in the aquatic environment could result in a low band rating for both the public and the fauna and flora, whereas for another radionuclide, the result might be a higher one for the fauna than for the public, or vice versa.

(132) At least two aspects of this approach need further consideration:

- that of restricting the advice to the effects of radiation on individual animals or plants, as is the case with human radiological protection; and
- that of restricting the advice in terms of radiation dose rates on fauna and flora to those that have particular observable effects.

With respect to the former, this is **not** to imply that it is the individual that is necessarily the object of protection in any particular exercise, but there are several reasons why it would be difficult to provide advice and recommendations at any other level. As noted in Chapter 2, there may be ethical or moral grounds or objections to mainly consider protection of the environment at the population or ecosystem level. Secondly, because a very large number of animals and plants are already afforded protection at the level of the individual, in international or national law with respect to some form of 'harm' arising from human activity, it would be inappropriate to attempt to provide advice that could not be used in such legal contexts. Thirdly, from a purely practical point of view, even to attempt to interpret the likely environmental consequences of many individuals, of any animal or plant receiving dose rates known to have effects at the level of the individual, in any particular circumstance, would require much more information of a non-radiobiological nature than could be incorporated into the provision of general radiation protection advice. But that is not to say that such information cannot be obtained, or applied to different circumstances, by national or international bodies. Thus, any approach developed by ICRP to estimate radiation doses to individual organisms and their associated biological effects must be amenable to be interfaced with the growing body of knowledge on ecosystem function to support assessments of contaminant effects on ecosystem integrity.

(133) Protection of non-human species may also have to be demonstrated, or taken into account, in many different circumstances. Thus the consequences of the presence of radionuclides in the environment may be managed by way of 'pollution control' legislation although, in some circumstances and for some countries, other legislative needs—such as those arising from 'nature conservation legislation—may predominate (Pentreath, 2003). Each of these may require different approaches, including those that are similar to toxicity-based or ecotoxicity-based approaches used in the management of other threats to the environment. They may require the local derivation of environmental standards—in terms of dose rates or radionuclide concentrations in particular environmental materials—to manage particular situations. Or they may simply require independent evaluations of the potential effects of radiation on the biological parameters of interest within any particular habitat or site. But these are decisions to be made at a national level. Additional and necessary advice and guidance would also be provided nationally via other fora. Nevertheless, it would greatly help the overall acceptability and interpretation of such decisions if they were all to be based upon—or shown to be derived from or related to—some system of reference methods, models, and databases.

(134) Setting out data in terms of radiation dose rates that were known to have particular radiation effects on different types of fauna and flora would appear to be the most appropriate *and transparent* format in which to provide general advice. This could be used to support legal frameworks at a national level that were already being drawn up in terms of 'dose-rate limits' for the environment, as in the USA, as a result of their assessment procedures, but it has to be borne in mind that other countries may not wish to pursue such a route, either in terms of legal interpretation, or in terms of using dose rates as the basis of any form of guidance or stricter form of legislative control.

6.3. Next steps for the Commission

(135) From all of the above, it is evident that the need to develop a common approach is urgent. It is also feasible. In recent years, a large amount of relevant work has been carried out by individuals, and by international and national organisations. There have been specific research programmes, specialist review groups, and interpretations of the large amount of radioecological information that has been gathered over the last 50 years. All of this work, plus the work of this ICRP Task Group, provides a basis for the development of a practical framework for the protection of both humans and non-human species. However, it first needs to be broadly introduced, developed, and hopefully accepted if some form of international consensus is to be achieved.

(136) That is not to say, as is the case with most of our experience with radiation protection, that there are no important data or knowledge gaps. One of the major gaps arises from an earlier lack of any systematic attempt to compile data specifically relevant to the protection of flora and fauna, both in the context of exposures and effects relevant to defined endpoints. Several time-limited initiatives are already taking place on this subject but ICRP could, in co-operation with others, immediately play a major role in compiling this information in a manner helpful to the development of a workable framework.

(137) Of particular urgency is the issue of the definition of relevant quantities and the selection of their associated units. A number of suggestions have recently been made (Pentreath, 1999; Kocher and Trabalka, 2000; Trivedi and Gentner, 2000; Thompson et al., 2003). Their principal differences are not simply in their choice of terminology but in the concepts that lie behind them. The choice of values and their basis needs detailed consideration; the scientific literature contains values that range over orders of magnitude, and the whole subject is in need of serious evaluation and further research.

(138) The Commission should therefore consider how best to resolve these and other issues in co-operation with other bodies. If it intends to develop this subject further, it would also need to consider how best to deal with such issues as the selection of primary reference fauna and flora, and their associated dose models and databases. These are both ethical/policy issues and radiation biology issues (except the first), but all will need a broader discussion with non-radiation bodies at a high level. It would also be useful to set out some clear 'divisions of responsibility' between ICRP, UNSCEAR, and IAEA in order to avoid duplication and the confusion of others.

7. CONCLUSIONS AND RECOMMENDATIONS

(139) As has been shown in this report, there is already a need for a broad international basis for evaluating and managing the actual and potential impact of radiation on the environment. This need arises in a number of ways, and from a variety of international and national circumstances. Several countries are already embarking upon their own programmes to address these issues, with the resultant risk of national divergence in both the approach and in the basic interpretation of radiation biology. Such fragmentation would be unhelpful, not only to the subject of environmental protection, but to the understanding and interpretation of the basic science of the interaction of radiation with living matter, be it human or non-human.

(140) The Task Group believes that there is a need to develop a framework for the assessment of radiation effects in non-human species. This need has not been driven by any particular concern over environmental radiation hazards. It rather needs to be developed to fill a conceptual gap in radiological protection and to clarify how the proposed framework can contribute to the attainment of society's goals of environmental protection by developing a protection policy based on scientific and ethical–philosophical principles.

(141) In contrast to the situation of human radiological protection, however, a considerable body of international and national legislation already exists that defines the general objectives of environmental protection and how it may be achieved. Guidance on the protection of non-human species from ionising radiation therefore has to be cast in such a way that it can meet existing needs. Such guidance should also be consistent with the scientific understanding of the relationships between radiation exposure, radiation dose, and radiation effects that is used as a basis for human radiological protection. It also needs to be articulated in such a way that it engenders confidence in decision makers and the public. This requires a transparent and coherent demonstration of the fact that the exposure of non-human species to radiation, in a variety of circumstances, can be adequately described and evaluated, and that the actual or potential consequences for them can be assessed and, if necessary, managed. In many cases, this needs to be done irrespective of whether similar assessments need to be made for human beings. This is frequently the situation with regards to the protection of the environment from other hazards, chemicals, and so on, and there is no compelling argument for radionuclides to be treated differently. The purpose of the reference fauna and flora approach is to enable different countries to assess and protect their environments in relation to their own national requirements, drawing on a common understanding at an international level of the effects of radiation on different types of animals and plants.

(142) The Commission is undoubtedly well placed to provide guidance on tools, methods, and data sets to assess radiation doses to biota, and estimates of risk that could be generally accepted and that could also synthesise protection of human and non-human species into a coherent framework. A prerequisite is that ICRP acknowledges the need for guidance by citizens, regulators, and implementers, and the need for awareness and concern in this matter. This would require that the current ICRP system for assessments be expanded in order to demonstrate explicitly that

impacts on non-human species can be assessed directly and a level of protection can be identified in a transparent manner. This, in turn, requires that a system for evaluation of risks to non-human species be developed. In the latter case, ICRP must also recognise the need to appreciate developments in other fields of environmental protection.

(143) The Commission should thus have a specific role to play. This role should be based on the need to provide over-arching policy and guidance by providing the sort of underpinning information that it provides for human radiological protection by way of recommendations and advice, supported by some key data sets and models. In order to develop a framework for the assessment of the impact of ionising radiation on non-human species, and protection against harmful effects of ionising radiation, ICRP therefore needs to revise its current system of protection, and particularly:

- develop a comprehensive approach to the study of the effects on, and protection of, all living matter with respect to the effects of ionising radiation;
- develop a system of radiological protection that includes protection of non-human species with a clear set of objectives and principles, and an agreed set of quantities and units applicable to all living things;
- interpret basic knowledge of radiation effects in species other than humans so that they can be used in an environmental context, for example, in setting criteria or benchmarks of protection at the appropriate level of hierarchy (individuals or populations);
- develop a small set of primary reference fauna and flora, plus their relevant databases, so that others can develop more area—and situation-specific numerical approaches to assessment and management of risks to non-human species;
- show its commitment to protection of non-human species and let this be reflected in the organisation of work and in the composition of experts; and
- plan regular reviews and revisions of this new system as new knowledge develops.

(144) The ICRP system of protection has evolved over time as new evidence has become available and as our understanding of underlying mechanisms has increased. Consequently, the Commission's risk estimates have been revised regularly, and substantial revisions have been made at intervals of about 10–15 years. It is therefore likely that any system designed for the radiological protection of the environment would also take time to develop, and similarly be subject to revision as new information is obtained and experience gained in putting it into practice.

REFERENCES

ACRP (2002) *Protection of Non-Human Biota From Ionizing Radiation.* Advisory Committee on Radiation Protection of the Canadian Nuclear Safety Commission (CNSC), INFO-0730, Ottawa, ON.

Amiro, B.D., Zach, R. (1993) A method to assess environmental acceptability of releases of radionuclides from nuclear facilities. Environ. Int. 19, 341–358.

Amiro, B.D. (1997) Radiological dose conversion factors for generic non-human biota used for screening potential radiological impacts. J. Environ. Radioact. 35, 37–51.

Barnthouse, L.W. (1995) *Effects of Ionizing Radiation on Terrestrial Plants and Animals: A Workshop Report.* Environmental Sciences Division, ORNL/TM-13141, Oak Ridge National Laboratory, Oak Ridge, Tennessee, USA. Sponsored by US Department of Energy.

Bergonié, J., Tribondeau, L. (1906) Interprétation de quelques résultats de la radiothérapie. Comptes Rendus Acad. Sci. 143, 983.

Bird, G.A., Thompson, P.A., Macdonald, C.R., et al. (2003) *Assessment of the Impact of radionuclide releases from Canadian nuclear facilities on non-human biota.* Third International Symposium on Protection of the Environment from Ionizing Radiation. (SPEIR 3). IAEA Vienna, Austria, pp. 241–247.

Brechnignac, F. (2001) Impact of radioactivity on the environment: problems, state of knowledge, and approaches for identification radioprotection criteria. Radioprotection 36, 511–535.

Brechnignac, F., (2002a) Environment vs. man radioprotection: the need for a new conceptual approach. In: Brechnignac, F. (Ed.), *The Radioecology-Ecotoxicology of Terrestrial and Estuarine Environments, vol. 37.* Radioprotection Colloques. EDP Sciences, Paris, pp. 161–166.

Brechnignac, F. (2002b) Protection of the environment: how to position radioprotection in an ecological risk assessment perspective. Sci. Total Environ. 307, 35–54.

CEAA (1992) *Canadian Environmental Assessment Act to Establish a Federal Environmental Assessment Process. Bill C-13.* Statutes of Canada 1992, Chapter 37.

CEC (1985) Council Directive 85/337/EEC of 27 June 1985 on the assessment of the effects of certain public and private projects on the environment. Official Journal L 175, 05/07/1985, P.0040–0048.

CEC (1992) Council Directive 92/43/EEC of 21 May 1992 on the conservation of natural habitats and of wild fauna and flora. Official Journal L 206, 22/07/1992, P.0007–0050.

CEC (1996a) Council Directive 96/29/Euratom of 13 May 1996 laying down basic safety standards for the protection of the health of workers and the general public against the dangers arising from ionizing radiation. Official Journal L 159, 29/06/1996, P.0001–0114.

CEC (1996b) Council Directive 96/61/EC of 24 September 1996 concerning integrated pollution prevention and control. Official Journal L 257, 10/10/1996, P.0026–0040.

CEC (2000) Council Directive 2000/60/EC establishing a framework for community action in the field of water policy. Official Journal 327, 22/12/2000, pp. 1–73.

CEPA (1999) *Canadian Environmental Protection Act, 1999.* Chapter C—15.31 (1999; c.33). Department of Justice, Ottawa, ON.

Clarke, R. (1999) Control of low-level radiation exposure: time for a change? J. Radiol. Prot. 19, 107–115.

Copplestone, D., Bielby, S., Jones, S.R., et al. (2001) Impact assessment of ionising radiation on wildlife. Environment Agency (UK) R&D Publ. 128, 222pp.

De Shalit, A.D. (2000) *The Environment: from Theory to Practice.* Oxford University Press, Oxford, UK.

Goodhead, D.T. (1994) Initial events in the cellular effects of ionising radiations: clustered damage in DNA. Int. J. Radiat. Biol. 65, 7–17.

Higley, K.A., Domotor, S.L., Antonio, E.J., et al. (2003a) Derivation of a screening methodology for evaluating radiation dose to aquatic and terrestrial biota. J. Environ. Radioact. 66, 41–59.

Higley, K.A., Domotor, S.L., Antonio, E.J. (2003b) A kinetic-allometric approach to predicting tissue concentrations for biota. J. Environ. Radioact. 66, 61–74.

Higley, K.A., Domotor, S.L., Antonio, E.J. (2003c) A probabilistic approach to obtaining limiting estimates of radionuclide concentrations in biota. J. Environ. Radioact. 66, 75–87.

Holm, L.-E., Hubbard, L., Larsson, C.-M., et al. (2002) Radiological protection of the environment from the Swedish point of view. J. Radiol. Prot. 22, 235–248.

IAEA (1976) *Effects of Ionizing Radiation on Aquatic Organisms and Ecosystems. Technical Reports Series No. 172.* IAEA, Vienna, Austria.
IAEA (1979) *A Methodology for Assessing Impacts of Radioactivity on Aquatic Ecosystems. Technical Reports Series No. 190.* IAEA, Vienna, Austria.
IAEA (1985) *Sediment Kds and Concentration Factors for Radionuclides in the Marine Environment. Technical Reports Series No. 247.* IAEA, Vienna, Austria.
IAEA (1988) *Assessing the Impact of Deep Sea Disposal of Low Level Radioactive Waste on Living Marine Resources. Technical Reports Series No. 288.* IAEA, Vienna, Austria.
IAEA (1992) *Effects of Ionizing Radiation on Plants and Animals at Levels Implied by Current Radiation Protection Standards. Technical Reports Series No. 332.* IAEA, Vienna, Austria.
IAEA (1995) *The Principles of Radioactive Waste Management. IAEA Safety Series No 111 F.* IAEA, Vienna, Austria.
IAEA (1996) *International Basic Safety Standards for Protection against Ionizing Radiation and for the Safety of Radiation Sources. Safety Series No. 115.* IAEA, Vienna, Austria.
IAEA (1999) *Protection of the Environment from the Effects of Ionizing Radiation, IAEA-TECDOC-1091.* IAEA, Vienna, Austria.
IAEA (2002a) *Ethical Considerations in Protecting the Environment from the Effects of Ionizing Radiation, IAEA-TECDOC-1270.* IAEA, Vienna, Austria.
IAEA (2002b) *Specialist Meeting on Protection of the Environment from the Effects of Ionizing Radiation: International Perspectives, IAEA-723-J9-SP-114.3.* IAEA, Vienna, Austria.
ICRP (1975). *Reference Man: Anatomical, Physiological and Metabolic Characteristics. ICRP Publication 23.* Pergamon Press, Oxford, UK.
ICRP (1977) Recommendations of the International Commission on Radiological Protection. ICRP Publication 26, Ann. ICRP 1 (3).
ICRP (1991) Recommendations of the International Commission on Radiological Protection. ICRP Publication 60, Ann. ICRP 21 (1–3).
ICRP (1994) Human respiratory tract model for radiological protection. ICRP Publication 66, Ann. ICRP 24 (1–3).
ICRP (1995) Basic anatomical and physiological data for use in radiological protection: the skeleton. ICRP Publication 70, Ann. ICRP 25 (2).
ICRP (1998a) Radiological protection policy for the disposal of radioactive waste. ICRP Publication 77, Ann. ICRP 27 (Suppl.).
ICRP (1998b) Radiation protection recommendations as applied to the disposal of long-lived solid radioactive waste. ICRP Publication 81, Ann. ICRP 28 (4).
ICRP (1999) Protection of the public in situations of prolonged radiation exposure. ICRP Publication 82, Ann. ICRP 29 (3–4).
ICRP (2001a) Doses to embryo and fetus from intakes of radionuclides by the mother. Ann. ICRP 31 (1–3).
ICRP (2001b) A report on progress towards new recommendations: a communication from the International Commission on Radiological Protection. J. Radiol. Prot. 21, 113–123.
ICRP (2002) Basic anatomical and physiological data for use in radiological protection: reference values. ICRP Publication 89, Ann. ICRP 32 (3-4).
IUCN (1980) *World Conservation Strategy 1980. Living Resources Conservation for Sustainable Development.* International Union for Conservation of Nature and Natural Resources (IUCN), Switzerland.
IUR (2000) *Doses and Effects in Non-human Systems.* International Union of Radioecology, Oslo, Norway.
IUR (2001) *Statement from the Consensus Conference on Protection of the Environment. Radiation Protection in the 21st Century: Ethical, Philosophical and Environmental Issues.* Norwegian Academy of Science and Letters, Oslo, Norway.
Jones, D.S. (2000) Radiological benchmarks for effects on aquatic biota at the Oak Ridge Reservation. Human Ecol. Risk Assess. 6, 789–807.
Jones, D.S., Domotor, S.L., Higley, K.A., et al. (2003) Principles and issues in radiological ecological risk assessment. J. Environ. Radioact. 66, 19–39.

Kocher, D.C., Trabalka, J.R. (2000) On the application of a radiation weighting factor for alpha particles in protection of non-human biota. Health Phys. 79, 407–411.
Linsley, G., Torres, C. (2001) The International Atomic Energy Agency's Programme on Biosphere Modelling and Assessment (BIOMASS). In: Proceedings of the Radioecology—Ecotoxicology of Continental and Estuarine Environments (ECORAD 2001). Radioprotection colloques, 37, Cl. pp. 133–139.
NCRP (1991) *Effects of Ionising Radiation on Aquatic Organisms. Report 109.* National Council on Radiation and Measurements, Bethesda, MD. 115pp.
NRC (1993) Setting Priorities for Land Conservation. US National Academy of Sciences Press, Washington, DC, USA
NRC (1996) *Understanding Risk in a Democracy.* National Academy Press, Washington, DC, USA.
OECD (1995) *The Environmental and Ethical Basis for Geological Disposal.* OECD-NEA, Paris, France.
OECD (2000) *A Critical Review of the System of Radiation Protection.* Committee on Radiation Protection and Public Health OECD-NEA, Paris.
OECD (2002) *Radiological Protection of the Environment—The Path Forward to a New Policy?* NEA Forum, Taormina, Italy, 12–14 February 2002.
OSPAR (1992) OSPAR Convention for the Protection of the Marine Environment of the North-East Atlantic. OSPAR Commission, London, UK.
OSPAR (1998) OSPAR Convention for the Protection of the Marine Environment of the North-East Atlantic. Ministerial Meeting of the OSPAR Commission, Sintra, 22–23 July 1998.
Pentreath, R.J. (1998) Radiological protection criteria for the natural environment. Radiot. Prot. Dosim 75, 175–179.
Pentreath, R.J. (1999) A system for radiological protection of the environment: some initial thoughts and ideas. J. Radiol. Prot. 19, 117–128.
Pentreath, R.J. (2002) Radiation protection of people and the environment: developing a common approach. J. Radiol. Prot. 22, 1–12.
Pentreath, R.J. (2003) *Evaluating the Effects of Ionizing Radiation Upon the Environment.* In: Third International Symposium on the Protection of the Environment from Ionizing Radiation (SPEIR 3). IAEA, Vienna, Austria.
Pentreath, R.J., Woodhead, D.S. (1988) Towards the development of criteria for the protection of marine fauna in relation to the disposal of radioactive wastes into the sea. Radiat. Protect. Nuclear Energy (IAEA, Vienna) 2, 213–243.
Pentreath, R.J., Woodhead, D.S. (2001) A system for protecting the environment from ionising radiation: selecting reference fauna and flora, and the possible dose models and environmental geometries that could be applied to them. Sci. Total Environ. 277, 33–43.
Pihet, P. (1998) Environmental and occupational dosimetry: an integrated approach to radiation protection covering radioecology, dosimetry and biological effects. A joint concerted action of EULEP, EURADOS and IUR in the Nuclear Fission Safety Programme of the European Commission. Newsletter EULEP EURADOS UIR 3, 9–10.
Rawls, J. (1971) *A Theory of Justice.* Harvard University Press, Cambridge, MA, USA.
Sazykina, T.G., Kryshev, I.I. (1999) Estimation of the control concentration of radionuclides in sea water considering health physics and radioecological criteria. Atom. Energ. 87, 302–307. (in Russian).
Shrader-Frechette, K., (1994) Risk and ethics. In: Lindell, B. (Ed.), *Radiation and Society.* IAEA, Vienna, Austria, pp. 167–182.
Stilling, P. (1999) *Ecology Theories and Applications, 3rd edn.* Prentice Hall, Upper Saddle River, NJ.
Strand, P., Brown, J.E., Woodhead, D.S., et al. (2000). *Delivering a System and Framework for the Protection of the Environment from Ionising Radiation.* Tenth International Congress of the IRPA, Hiroshima, Japan, 14–19 May 2000, P-2a-116, 5 pp.
Strand, P., Larsson, C.-M., (2001) Delivering a framework for the protection of the environment from ionising radiation. In: Brechignac, F., Howard, B.J. (Eds.), *Radioactive Pollutants. Impact on the Environment.* EDP Sciences, Les Ulis, France.
Suter II, G.W. (1999) Developing conceptual models for complex ecological risk assessments. Hum. Ecol. Risk Assess. 5, 375–396.
Thompson, P.A., Macdonald, C.R., Harrison, F. (2003) *Recommended RBE Weighting Factor for the*

Ecological Risk Assessment of Alpha-emitting Radionuclides. In: Third International Symposium on Protection of the Environment from Ionizing Radiation (SPEIR 3). IAEA, Vienna, Austria.
Trivedi, A., Gentner, N.E. (2000). *Ecodosimetry Weighting Factor (er) for Non-human Biota*. Tenth International Congress of the IRPA, 14–19 May 2000, Hiroshima, Japan, 2000, P-2a-114, 8pp.
UN (1945) *Charter of the United Nations*. United Nations, San Francisco, USA.
UN (1972) *Declaration of the United Nations Conference on the Human Environment 21st Plenary Meeting, The United Nations Conference on the Human Environment, Sockholm*. United Nations, New York, NY.
UN (1992) *Rio Declaration on Environment and Development, Rio de Janeiro*. United Nations, New York, NY.
UNSCEAR (1986) *Sources and Effects of Ionizing Radiation. Report to the General Assembly with Scientific Annexes*. United Nations, New York, NY, USA.
UNSCEAR (1996) *Sources and Effects of Ionizing Radiation. Report to the General Assembly with Scientific Annex*. United Nations, New York, NY, USA.
UNSCEAR (2000) *Sources and Effects of Ionizing Radiation. Report to the General Assembly with Scientific Annexes*. United Nations, New York, NY, USA.
UNSCEAR (2001) *Hereditary Effects of Radiation. Report to the General Assembly with Scientific Annex*. United Nations, New York, NY, USA.
USDOE (1993) *Radiation Protection of the Public and the Environment. DOE Order 5400.5*. US Department of Energy, Washington, DC, USA.
USDOE (1996) Radiation protection of the public and the environment. Federal Register 61 (36), 6799–6801.
USDOE (2002) *A Graded Approach for Evaluating Radiation Doses to Aquatic and Terrestrial Biota, Technical Standard DOE-STD-1153-2002*. US Department of Energy, Washington, DC, USA.
Van der Stricht, E., Kirchmann, R. (2001) Radioecology: Radioactivity and Ecosystems. Fortemps, Liège, Belgium.
Webster, S. (2000). Environmental impact assessments and geological repositories—a model process. International Conference of the Safety of Radioactive Waste Management, Córdoba, Spain, 13–17 March, International Atomic Energy Agency, pp. 294–297, IAEA-CN-78.
Waste Convention (1997). Joint Convention on the Safety of Spent Nuclear Fuel Management and on the Safety of Radioactive Waste Management. INFCIRC1546. IAEA, Vienna, Austria.
Woodhead, D.S. (1973) The radiation dose received by plaice (Pleuronectes platessa) from the waste discharged into the north-east Irish sea from the fuel reprocessing plant at Windscale. Health Phys. 25, 115–121.
Woodhead, D.S. (1979) Methods of dosimetry for aquatic organisms. In: *Methodology for Assessing Impacts of Radioactivity in Aquatic Ecosystem, IAEA Technical Report 190*. IAEA, Vienna, Austria, pp. 43–96.
World Commission on Environment and Developmemt (1987) Our Common future (The Bruntland Report). Oxford University Press, Oxford, UK.

FSC
www.fsc.org
MIX
Papier | Fördert
gute Waldnutzung
FSC® C083411

Kolibri 360 GmbH
Ferdinand-Jühlke-Straße 7
99095 Erfurt, Deutschland
produktsicherheit@kolibri360.de